The Dyer's Companion

ELIJAH BEMISS

THIRD ENLARGED EDITION

With a
New Introduction by
RITA J. ADROSKO
Curator, Division of Textiles
Smithsonian Institution

DOVER PUBLICATIONS, INC.
NEW YORK

Published in Canada by General Publishing Company, Ltd., 30 Lesmill Road, Don Mills, Toronto, Ontario.

Published in the United Kingdom by Constable and Company, Ltd.

This Dover edition, first published in 1973, is an unabridged republication of the second (1815) edition as published by Evert Duyckinck in New York. This reprint edition also contains a new introduction by Rita J. Adrosko, and two Appendices reprinted from *Natural Dyes and Home Dyeing* by Rita J. Adrosko (Dover Publications, New York, 1971).

International Standard Book Number: 0-486-20601-7
Library of Congress Catalog Card Number: 73-77377

Manufactured in the United States of America
Dover Publications, Inc.
180 Varick Street
New York, N. Y. 10014

INTRODUCTION
TO THE DOVER EDITION

Only a few facts are known about Elijah Bemiss' life. By the time he wrote the first (1806) edition of *The Dyer's Companion,* he was a 45-year-old veteran of the American Revolution. Sometime between the end of his war service and 1806 he had moved from Massachusetts where he was born to Waterford, Connecticut where he established a clothier's business.

We learn of Bemiss' location and trade from an advertisement Bemiss placed in the October 5, 1803 *Connecticut Gazette* in which he

> begs leave to acquaint his old customers, and the public in general, that he continues to execute with neatness and dispatch all kinds of business in the *Clothiers's* line, at his mill on Alewife Brook . . . where constant attendance will be given and the smallest favours gratefully acknowledged . . .

The first edition of *The Dyer's Companion* was the second known dye manual printed in the United States. The first, Asa Ellis' *Country Dyer's Assistant* was printed in 1798. Both books contained information especially applicable to the needs of American dyers, unlike some later publications which were simply reprints of English dyebooks or translations of French works.

The success of Bemiss' book is attested by the fact that the first edition, privately printed in New Haven, was followed by the commercial publication of the greatly expanded 1815 edition reprinted herein. This second edition of *The Dyer's Companion* was published by Evert Duyckinck, a New York printer well known for issuing standard American works. Without doubt he was aware that the many small dye workshops cropping up all over the northeast provided a ready market for a practical dye manual. Professional dyers, who served the needs of a growing American textile industry, needed the latest available information and the revised edition of this book satisfied this need.

The 1815 edition boasts a few new recipes and slight revisions of some old ones. More importantly, an all-new 187-page "Appendix" was added. This contains results of the author's further experience and selections from contemporary European authors, chosen for American dyers' use. The last section of the work is a collection of miscellaneous "useful receipts," including inks and varnishes, wines, cancer cures and insecticides. Although much of this material has little practical value, it is retained as a matter of curiosity and historical interest.

The original intention of this book, as stated by its author was "not so much to please others . . . but to communicate . . . good wishes for improvements to my brethren the dyers, and to show my willingness to help to perfect one of the most useful arts in the world." Today most readers can appreciate *The Dyer's Companion* simply as a historical document of the period. Although its major portion is devoted to dye recipes, there is much in this little volume that reveals the author's attitudes toward his work, his fellow countrymen and his country.

Contemporary students of natural dyes who enjoy the benefits of modern chemical knowledge will be amused by Bemiss' at times fanciful explanations of the chemistry of dyeing. But one should not discount the author's practical knowledge of dye procedures based on years of experience and observation. This book offers many worthwhile suggestions for handling dye materials and combining them.

Before attempting to use any of these recipes, certain problems inherent in using any early dye recipes should be recognized. Even in the early nineteenth century this book was intended for use by professionals, with recipes usually calculated for 20 or more yards of cloth. Since most contemporary craftsmen find one pound of textile material most convenient to handle in a home dyeing situation, obvious adjustments in proportions of ingredients would be necessary. This involves more than a simple reduction, however, for Mr. Bemiss' recipes deal with yards of cloth, rather than pounds of textile material. Although nineteenth-century clothiers may have had some knowledge of weight per yard of cloth, the contemporary dyer has no way of obtaining this information.

Since the state of early nineteenth-century chemical manufacture was rather primitive by modern standards, dye chemicals used then were considerably less pure than those available today. Thus modern chemicals' potency would differ enough so proportions again would have to be adjusted. Lack of standardization of modern natural dyestuffs would make the quantities stated merely rough estimates. Just as there were various grades of early dyestuffs, modern natural dyes, such as madder, may differ greatly from supplier to supplier, depending on whether or not they contain twigs or other natural adulterants which affect their potency.

Finally, there is no way of judging the standards by which early dyers measured the success of a dye job. While clear, even colors have been the goal of dyers of all eras, other standards for judging success of a dye job could have changed over the years. Modern dyers usually strive for soft, non-compacted products; it is quite possible that certain early woollens, for example, might have been considered desirable if solid and firm. Thus even if one succeeded in surmounting all other obstacles, following Bemiss' recipes too literally could lead to results that could be considered poor by present-day standards.

If you wish to experiment with these recipes, it is advisable to first select a reliable modern book on the subject, which offers "how-to-do-it" instructions. Once you have tried a few recipes and feel you understand certain basic features, such as the usual proportions of mordants, dyestuffs and dye chemicals, and how and when they are used, you are ready to experiment with Bemiss' recipes.

The present edition contains two new appendices taken from *Natural Dyes and Home Dyeing* by Rita J. Adrosko (Dover, 1971), as follows: "Common Names of Chemicals used in Dyeing" and "Dyes Occasionally Mentioned in Dyers' Manuals Printed in America." The first of these appendices provides the modern chemical terminology for many of the substances which Bemiss refers to by their nineteenth-century common names. The other appendix is of further interest for the practical application of this volume, supplying the names of additional nineteenth-century American plants used in dyeing.

Certainly *The Dyer's Companion* offers students of dyeing and craftsmen with a love of old methods and previous dyeing experience many ideas worth pursuing.

RITA J. ADROSKO

Washington, D. C.
1973

A Few Sources of Natural Dyes

Dharama Trading Company
P. O. Box 1288
Berkeley, California

Dominion Herb Distributors Inc.
61 St. Catherine St. West
Montreal, Canada

Worldwide Herbs
Box 378
Streetsville, Ontario, Canada

D. Bailey
15 Dulton Street
Bankstown, N.S.W. 2200
Australia

Straw into Gold
5550-H College Avenue
Oakland, California 94618

Earth Guild
Hudson Street
Cambridge, Massachusetts

Note: This list, compiled in fall, 1972 is subject to change without notice. Advertisements in craft publications such as *Craft Horizons, Handweaver and Craftsman* and *Shuttle, Spindle and Dyepot* are the best sources of up-to-date information on natural dye sources. Chemical houses and janitorial supply houses also sell dye chemicals such as mordants.

PREFACE.

THE design of "*The Dyer's Companion,*" is to furnish an easy and uniform system of dying for the use of practitioners, and those who wish to be benefitted by that and other arts introduced in this work. During an employment of several years in the clothier's business, I had to combat with many difficulties for the want of an assistant of this kind; and I am well persuaded the greater part of my fellow-functioners have laboured under the same embarrassments, as there has not been to my knowledge, any book of this nature ever before published in the United States—a work which I humbly conceive will not only be serviceable to the practitioners, but to the country at large.

The author's attempt to improve the useful arts, and to promote manufactures, he hopes will meet the approbation and encouragement of his fellow-citizens; and that the plainness of his plan, will be excused, as he is an unlettered country dyer. His long practise in dying and dressing cloth, &c. has given him great opportunity for making improvements therein. These arts admit of still greater improvement, if proper attention is paid to recording and securing our discoveries; but otherwise it must be expected that they will remain with us in a state of infancy.

The art of dying is still far from having arrived at a state of perfection even in Europe, and probably will not in our age. This consideration ought not to discourage us, but to increase our ambition; for it must be acknowledged that great improvements have been made and are still making in this country.

Those to whom the author is in the smallest degree indebted for promoting the usefulness of this work, will please to accept his thanks; their future favors are requested, with a hope that we may continue to live in brotherly love.

PREFACE.

By contributing our mutual aid towards gaining and supporting our independence of Great-Britain, and other foreign countries, to whom in arts and manufactures we have too long bowed the knee; we shall promote our own interests and our country's welfare and glory.

In the *First Part* it is attempted to have the Receipts for dying woollen, silk, cotton and linen goods, arranged in the best order; which is followed by directions for the management of colouring, &c. The different operations of dye-stuff are then attempted to be shewn, together with directions for dressing cloth; closing with some observations on the present situation of our business.

The *Second Part* contains several useful arts and discoveries, collected from various sources, which will be found to be extremely beneficial to the public in general.

The author having for several years practised in the greatest part of the arts inserted in this work, pledges himself for the truth of his assertions. He has endeavoured to use the plainest language, and to point out every part of the processes, so that no one should be disappointed who attempts to follow his directions.

Many master mechanicks refuse to give receipts to their apprentices unless they will pay for them, and at a high price. There are many receipts in this book, which, to the personal knowledge of the author, have been sold for twenty and thirty dollars each; and the purchaser prohibited from communicating the receipt to any other person. By this means, useful discoveries are sometimes wholly lost; and our improvement in arts and manufactures make but slow progress.

Should this attempt meet with reasonable encouragement the work will be enlarged and amended, in future editions, as the author may find time and means for the purpose.

INDEX.

PART FIRST.

INDEX.

INDEX.

PART SECOND.

INDEX.

THE

DYER'S COMPANION.

RECEIPTS, &c.

1. *To set a blue Vat of twelve Barrels.*

FOR a vat of twelve barrels; fill the vat about half full of water, scalding hot; dissolve eight pounds of potash in eight gallons of warm water; fill the copper with water; add one half of the potash lie, with five pounds of madder, and four quarts of wheat bran; heat this with a moderate fire, nearly to boiling heat, often stirring it—turn this into the vat. Take five pounds of indigo, wet it with one gallon of the potash lie, and grind it well: then fill your copper with water, and add the remainder of your potash lie, when cool, (being careful in pouring it off, as the sediment is injurious to the dye); add this compound of indigo, &c. and four pounds of woad; stir this continually over a moderate fire, until it boils; then turn it into the vat, and stir, rake or plunge well, until well mixed together; cover it close and let it stand two hours; then add four ounces of borax, rake well, and let it stand twelve hours.

If it does not come to work, then take two quarts of unslacked lime, and six quarts of water, putting them into a vessel proper for the purpose, and stirring well; after standing till well settled, take the lie of the lime, and rake again, cover close, and let it stand two hours. The symptoms of the dye being fit to work, may be known by the rising of a fine copper-coloured scum, on top of the dye, and likewise, a fine froth rising, called the head ; your dye will look green, and your cloth dipt in it, before it comes to the air, will look green also.

Form of a Vat and other Utensils necessary for Blue Dying.

1st. The *Vat* ought to be made of pine plank, at least two inches thick : it should be five feet long, and the width sufficient for containing the quantity required ; the largest end down, and about three feet in the ground ; hooped with large iron hoops as far as it stands in the ground ; and all above ground covered with wooden hoops ; the top covered tight with a thick cover so as to exclude the cold air. A small lid should be made to open and shut at pleasure for the purpose of admitting the dye into the vat, stirring, raking, &c. It is absolutely necessary to cover close, so as to confine the heat and steam from the time you begin to empty your liquor, until your vat is full. The liquor should be conveyed from the copper to the vat by a spout or trunk, and after stirring, be immediately covered close.

2d. The *Rake* is of an oval form, with a handle through the middle, of sufficient length to reach the bottom of the vat with ease.

3d. The *Screen* or *Raddle*, to prevent the goods from sinking upon the sediment. This utensil is placed about ten or twelve inches from the bottom of the vat. It should be as large as the top of the vat will admit, and filled with netting or splinters ; it should be hung by three cords from the top, so as to be easily taken out when necessary, and a weight in the middle sufficient to keep it down.

4th. The *Cross-Bar*, or stick across the vat. This should be about one inch in diameter, and placed about six inches from the top, and across the middle of the vat.

5th. The *Handlers*, *Claws* or *Hooks*, are for managing the cloth in the dye, (for no air must come to the cloth while in the dye). The claws are made with wooden handles; the hooks of iron in an oval form, half round, and notches in the hooks like saw teeth, for the purpose of catching hold of the cloth.

To fit Cloths for Dying.

In the first place scour the grease well out of the cloths. Take about thirty yards of cloth to a fold or draft, having prepared, in your copper, about two barrels of water, with four ounces of pearlash therein; in this liquor run and prepare your cloth for the vat about eight or ten minutes; then roll it out and let it drain. Then fold it up smooth on the side of the vat, that it may go in open; toss the end over the cross-bar, and let a person on the other side with his handlers be ready to poke it down, and let it be done quick and lively. When the cloth is all in the vat, take the other end back again, by pulling it hand over hand, very lively, till you arrive at the other. Then shift sides, and manage in this manner till ready for taking out; which will be in ten or twelve minutes, if the dye is ripe and hot. But judgment must be used in this case; when the dye is weak and cool, it is necessary to keep the cloth in an hour or more:

In taking the cloth out of the vat, it is necessary to use dispatch. The utensils for this purpose are two crooked irons passed just above the vat, so that two men may put the cloth thereon, as taken out of the vat; then a windlass for the purpose of wringing the cloth as dry as conveniently can be done. Hang your cloth then in the open air, till it is perfectly cool. At the same time, if you have more cloth, prepare it as described before in the copper of pearlash water. This process must be observed every time the cloth is dipped in the vat. Two dippings are commonly sufficient for colouring the first time; then air and rince, and this will be a pretty good blue—and full and manage as you do cloths to prepare them for colouring. However, your dye must not be crowded too fast at first.

If you find your dye does not colour fast enough, cover and rake, and let it stand an hour or two; being careful to keep the vat covered, excepting when the cloth is in: work the dye till it is cool, then heat it again. If all your cloths are not coloured for fulling; heat your dye again in the copper or other utensil, nearly to boiling heat, then turn it into the vat and cover it up; add two pounds of pearlash, rake well, and let it stand ten or twelve hours; then rake it, and let it stand two hours, when it will be fit for work. Let the dye be worked as long as it will colour well; then manage as before until the dye is reduced. Recruit as before in setting, and manage in the same manner till your cloths are all coloured. Only omit two pounds of potash and one pound of indigo out of the quantity; and the dye must stand to come to work, which will probably be sooner than at first; caution must be used about working it too soon.

The cloths when fulled and prepared for colouring, must be managed as at first, and run till they suit. After you have done colouring, open your vat, rake well, and give the dye all the air you can. Let it stand, and it may be kept good for many years, if rightly managed: After it has been recruited several times, it will be necessary to dip off the dye carefully so as not to disturb the sediment or lees, and throw the lees away. When the dye has been standing a long time, it is necessary to throw away the lees, for they will have a tendency to injure the dye, and the colour will not be so bright if they remain in the vat. The dye will not come to work so soon as if the sediment had remained in the vat, and it ought not to be disturbed excepting when it is necessary to dispense with some of the lees.

The dyer being careful to manage according to these directions, will have the best mode of dying cloth blue, known by me.

To color yarn or wool in this dye, the yarn must be hung loose in the dye, and the wool be put loose into a nett and then immersed.

When the goods are dyed, have them immediately rinced in clear water; when dryed, take twelve gallons of warm water to one pound of hard soap dissolved, and one pint of beef gall; wet the cloth with this, and let it run in the mill eight or ten minutes, then rince it with fair water till perfectly clean, and it will prevent the goods from crocks, &c. if the color is not struck through the cloth and cuts light in the middle, to 20 yards take half pint of color, put in your copper of boiling hot water, run one hour, and rince well.

2d. ANOTHER METHOD FOR BLUE.

The best to dye Yarn or Wool.

TO set a tub of 6 gallons, take five gallons of good old sig, to which add 2 gills of spirits, half a pound of good indigo made fine; put it in a bag, wet it and rub it out in the dye, then add two ounces of pearlash, and 2 ounces of good madder; stir and mix it all together, let it stand 24 hours; then add half a pint of wheat bran, stir it up till well mixed together, let it stand 24 hours longer, and if your dye does not come to work by this time, stir it as often as once in two or three hours, but do not apply your goods before your copper scum and froth rises, and the dye looks greenish when dropping, and your yarn or wool looks greenish when applied to the dye, which are symptoms that your dye is in good order for use; but you must be cautious not to crowd your dye too full, for many blue dyes are destroyed in this way. Be careful also about reducing your dye too low; always keep indigo in the bag, rubbing it out when necessary; and you need not stop your dye to recruit it after it has come to work; but make your additions when you take your goods out, as you find it necessary. Wring out the goods, stir your dye well together, cover it close, and place it where it will keep lukewarm. It will not dye so quick as the other dye, but it will make a superior blue. It is commonly from two to three days in colouring for a deep blue.

N. B. The yarn or wool should be wet in warm sig, before it is put in the dye, and the tub covered close, &c.

3d. *ANOTHER METHOD FOR BLUE.*

TAKE half a pail full of good ashes, two quarts of stone lime, and as much sig as to run through three gallons of liquor; add two ounces of good indigo made fine, four ounces of good madder, and half a pint of wheat bran; stir and mix it well together, let it stand two days, then stir it up, and put in half a pint of good emptines. Let it stand 24 hours, and your dye will be fit for work.

Directions to be observed in common Colouring.

EVERY person that understands his business knows what utensils are necessary for the business in colouring; however, I will give a brief description of those commonly used.

The first thing necessary is the copper kettle; I say *copper* kettle, because it is most commonly used in all hot dyes, and all hot dyes may be coloured in the copper, and I shall mention no other in the following receipts. Block tin or brass, are better for red and yellow, than the copper; and iron the best for black or green; but this I leave to the discretion of those in practice. The size ought to be from two to four barrels, according as your business requires. In setting the kettle, reference should be had to convenience of heating and working.

The *Reel*, as it is commonly called, which is used for managing the cloth in the dye, is conducted over and over in the dye, being turned by a *wench;* and the cloth is poked down and spread open by a stick about three feet long. The cloth always should be tended lively when in the dye. (The time the cloth is to be in these dyes, will hereafter be described.)

When the cloth has been a sufficient time in the dye, then reel or wind it up; let it drain a few minutes, then take it out in the open air, and spread it till perfectly cool; and this must be the management every time the cloth is dipped. Never add any dye-stuff or water when the cloth is in the dye; but when added, stir and mix the dye well together before the cloth is put in. The cloths should be perfectly cool to prevent their spotting, and for the brightness of colours have the kettle well cleaned. To clean a copper, the most common form I practise, is to rince the dye well off, then take some ashes and a swab, and rub it well and rince it clean, and it will answer for most colours. But if it does not appear bright enough, then take half a gill of oil of vitriol, and rub in the same manner as before; rince clean, &c.

To clean a Copper.

TAKE four ounces of allum, two quarts of vinegar, and two ounces of oil of vitriol; put them all together, heat them boiling hot, and put them into your kettle; wash it well with a swab, rince it with water clean, and it will be fit for any dyes.

A GENERAL RULE.

I SHALL lay it down as a general rule, to take 20 yds. or 16 lbs. weight for the quantity of cloth, for which to proportion the dye-stuff. However, any quantity of cloth or goods may be coloured by the following receipts; only in the like proportion as before mentioned: and another thing is to be observed, the different states of the dyes, by giving all your goods an equal chance in the dye; for most of colours the dye is good for nothing for that colour after the colour is done.

4th FOR BLUE.

TO 20 yds. of fulled cloth, take four pounds of good logwood chips; fill your copper with fair water, add the logwood, and boil well till the strength is out; then add one pound of good madder and one pound of allum; let it simmer together fifteen minutes, but not boil, (for the madder ought never to boil (run your cloth twenty or thirty minutes, roll out and air it; let the dye simmer a few minutes, then run it again as before, with the heat of the dye increasing, about thirty minutes: air it, and the cloth will then appear of a purple cast or shade. Then take two ounces of verdigrease pulverized fine; then take one pint of sig; put them into a proper vessel, and simmer them together with constant stirring, till well mixed and dissolved; then add this to your dye, with two gallons of sig, and two ounces of blue vitriol; boil them moderately together about 15 minutes, then stop your dye from boiling, and stir well together, then run your cloth about thirty minutes: run in this manner till the colour suits, and you will have a fine blue, but it will not be so durable as Indigo blue.

5th. FOR NAVY BLUE.

TO twenty yards of fulled cloth; fill your copper with fair water, heat it boiling hot, take two pounds of copperas, half a pound of allum, a quarter of a pound of argal, or red tartar—pulverize these together, and put this compound into the boiling water—skim your dye, stop its boiling, run your cloth twenty or thirty minutes, air and run it again, as before, twenty minutes, air and rince it in water; shift your liquor from the copper, rince your copper, fill it with fair water, then add four pound of good logwood chips, boil well twenty minutes, then slacken your fire and add an half pound of good madder; let it simmer fifteen minutes—together with one ounce verdigrease made fine, as described in receipt fourth, with sig, &c. then take one gallon of sig and add with the rest to the dye, stir them well together, till the dye is well mixed ; run your cloth again in this dye thirty minutes, air it and add two ounces of pearl-ash and run it again, with the dye well mixed together—handle in this manner, till your colour pleases. This will be a good blue, rather preferable to receipt No. 4.

6th. PRUSSIAN BLUE.

Compound, or Chymic —This compound or blueing is made thus: Take one pound of good flotong indigo pulverized, four pounds of oil of vitriol, and two ounces of fine salt—put this in a stone pot (or some earthen vessel) that will contain six times the quantity of this compound, or it will be liable to rise and run over. First put in the vitriol, then the indigo, then the salt ; stir this continually one hour, or till it gets pretty well settled and cool—for it will boil and foment in a terrible manner. Let it stand four days or a week, covered close, stirring it now and then, as is most convenient.

7th. ANOTHER METHOD FOR BLUEING, OR COMPOUND.

TAKE one pound of common good indigo, six pounds of oil of vitriol, half a pound of stone lime—put these together (as described before) in the pot and stir it—This will be fit to use in forty eight hours. I have mixed it without either lime or salt ; but it requires more stirring and longer standing before it is fit for use. This compound is used for dying Prussian blue, green and many other colours.

8th PRUSSIAN BLUE.

FILL your copper with fair water, heat it nearly boiling hot, then add of your blueing (as is before mentioned) a little, and stir it well with the water, run your cloth, roll out, air, and add of your compound by little and little, till your colour pleases.—You may make in this dye, any shade you wish of this kind of blue, and very bright.

9th. FOR GREEN.

TO twenty yards of cloth, take six pound of fustick chips and boil them well, then add one quarter pound of allum, run your cloth till it is a good yellow, then add of your blueing* about half a gill at a time, stir and mix it well together in the dye, run your cloth with a hot fire fifteen or twenty minutes, then air and add a little of your blueing and run again in the same manner as before, and add of your blueing, little by little, till your colour suits.

If you have a considerable quantity of cloth to colour, it will be necessary to boil your fustick till your dye is strong; then put it in a tub for the convenience of dipping it off as it is wanted to mix with the bluing. The quantity of yellow dye to be dipped off, must be left to the discretion of the dyer, according to the quantity of cloth in colouring; let the chips remain in

* This compound of vitriol and indigo, is known by the blueing *chymick* or *saxon pot*.

the kettle, and fill your copper with water, boil again, and yellow your cloth till a good yellow, by adding allum every dipping—then take the chips out of the dye, then add of your blueing run all your clothes, then add of your blueing and yellow dye, having your dye hot and well mixed together—run your cloth, and add of your compound and yellow dye, by little and little, well mixed and stirred together; and if the colour does not appear bright enough, frequently add a little allum, keep it in much longer, and this will give lustre to your colour. This is the best method of dying a bright green, I believe in the world, or the best I ever knew.

Green requires the judgment of the dyer to prevent one colour from overrunning the other, otherwise the colour will appear dull, and never can be made bright. But follow the receipt with care and judgment, and you will have a very fine green.

10*th*. *FOR GREEN.*

TO twenty yards of cloth, take five pounds of good fustick chips, boil well, then add two ounces of allum, run your cloth till a good yellow; then add of your blueing half a pound, run your cloth twenty or thirty minutes, then air, and add a little copperas and a little logwood; let it boil a few minutes, run again and handle till your colour pleases.

11*th*. *FOR GREEN.*

TO twenty yards of cloth take four pounds of fustick chips, boil well, then add two ounces of pearlash, one ounce of allum, one ounce of aquafortis—let it boil, stir and mix it well together, then run your cloth till a good yellow; air, and add of your blueing, mix well with your dye, run your cloth, and add of your blueing by little and little, till your colour pleases.

12*th.* *FOR GREEN.*

TO twenty yards of cloth, take four quarts of wheat bran, wet it with vinegar, let it stand twelve hours; fill your copper with fair water, put your bran in a bag and let it boil in the water one hour, take it out, let it drain, and squeeze it dry as you can; then add two ounces of argal,* made fine, and one ounce of allum; boil well, run your cloth forty minutes, boiling; then air and rince, shift your liquor from your copper, rince and fill with fair water; then add four pounds of fustick chips, boil well till the strength is well out, then add a little allum, and run your cloth thirty minutes more; then add gradually, as much blueing as is necessary, and sadden with a little copperas.

If the colour is not bright enough, shift your dye from your copper, and fill with fair water; heat it nearly to boiling heat, add a little blueing, and handle till your colour pleases.

13*th* *FOR GREEN.*

TO twenty yards of cloth, take five pounds of fustick chips, and boil well; then add two ounces of allum, and six ounces of compound or blueing—half of your blueing at a time; run your cloth thirty minutes, then add the rest of your blueing together with yellow dye and a lit-

* This is called by some, *Crude*, or *Red Tartar*.

tle allum; run again as before; then add two ounces of blue vitriol, boil well, and handle till your colour pleases.

N. B. These green dyes are worth saving as they are useful in many dyes, especially for bottle green in the first beginning.

14th. *FOR BOTTLE GREEN.*

TO twenty yards of cloth, take three pounds of fustick chips, boil well, then add two ounces of allum and your blueing; stir and mix them well together, then run your cloth thirty minutes, air and run again till you have it a good deep green; then add two pounds of logwood, boil well, take one quarter of a pound of verdigrease, pulverize it, and put in a proper vessel with one pint of vinegar; let it simmer together with constant stirring, till all dissolved; then add it to the dye, stir and mix it well together, run your cloth with your dye hot, thirty or forty minutes; then air and sadden with copperas, till the colour is dark enough.

If your green goes off, shift your dye from your copper, clean it well, rince your cloth well, fill your kettle with fair water, heat it boiling hot, and add blueing by degrees till your colour pleases.

15th. *FOR BOTTLE GREEN.*

FOR twenty yards of cloth, fill your copper with fair water, heat it boiling hot; take half a pound of blue vitriol, and let it dissolve in the water; run your cloth 30 minutes, air and run again as before; then add three pounds of good logwood chips and two pounds of fustick, and boil well; run your cloth, and handle till your colour pleases; and you will have a fine bottle green, but it is more liable to fade than the other, which will hold equal to a blue.

Or this, take one pound blue vitriol, heat your copper with fair water, near boiling hot, run your cloth, then air and run again as before; then air, run and shift your liquor, then add 6 pound fustick and 4 pound logwood chips, boil well and run again as above, &c.

16*th.* *FOR OLIVE GREEN.*

TO twenty yards of cloth, take six pounds of fustick, boil well, then add a quarter of a pound of allum, and a quarter of a pound of blueing; run your cloth one hour, then add half a bushel of butternut bark; let it boil moderately till the strength is well out; run your cloth 30 minutes, air, and run again; then add one quarter of a pound of copperas, and handle till your colour pleases.

When I have any bright green dye, as in receipt No. 9, I use it as a preparation for the olive green.

17*th.* *FOR YELLOW.*

TO twenty yards of cloth, take a quarter of a pound of aquafortis, and as much pewter or block tin as the aquafortis will dissolve; (first pouring the pewter in a melted state into water;) fill your copper with fair water, heat boiling hot; then add the compound of aquafortis, &c. with six ounces of argal, and half a pound of allum; boil well, run your cloth boiling forty minutes; then air and rince, and shift your liquor from your copper; fill with fair water, then take four pounds of good fustick, and a quarter of a pound of turmerick, boil well, and add half a pound of allum; run your cloth thirty minutes, and handle till your colour pleases.

18*th.* *FOR YELLOW.*

TO twenty yards of cloth, take one pound of allum, fill your copper with fair water, heat boiling hot, run your cloth boiling, three quarters of an hour; air, rince and shift your liquor from your copper; rince and fill with fair water; add six pounds of good fustick, boil well, then add a quarter of a pound of allum, and two ounces of aquafortis killed with pewter as described in receipt No. 17; stir and mix it well together with your dye; run your cloth and handle till your colour suits your fancy.

The dyer must be exceeding careful in these yellow dyes, that his copper utensils and cloth are all clean; for the yellow dyes are very easily spoiled. It also requires great care about handling the cloths, that you do not touch them against any thing that will spot them, for that is not very easily mended.

N. B. The aquafortis must be put in a sound earthen or glass vessel, to contain much more than the quantity of aquafortis; for it will boil and fly, and appear to be red hot when you put in the pewter or block tin; and it must be fed as long as it will dissolve it. Then let it stand till cold; and stopped with wax or glass ston-per and it will keep good for work, then apply it to the dye. This is the way that aquafortis must be used, except otherwise directed. Remember the pewter or block tin must be melted and thrown into water, and it will dissolve the better, &c.

19th. *BUFF YELLOW.*

TO twenty yards of cloth, take four pounds of good fustick, boil well; then add a quarter of a pound of the best madder and six ounces of allum; let it simmer together, but not boil, (for the madder must not boil, but be near boiling) run your cloth, and handle till your colour pleases.

N. B. The yellow dye (after you have done dying your yellow,) may be useful to all colours that have yellow in them; for green, olive, &c.

20th. *TO TAKE THE COLOUR OUT OF CLOTH.*

TO twenty yards of cloth, take two pounds of red tartar, four pounds of allum, three quarters of a pound of cream of tartar, one pound of white argal or tartar; pulverize and mix them together; fill your copper with fair water, heat boiling hot; then add your compound, let it boil, run your cloth one hour boiling; and this will completely destroy almost any colour or colours.

21st. *FOR YELLOW.*

AFTER you have taken the colour out. The cloth must be well rinced in water. For twenty yards of cloth fill your copper with fair water, then add two pounds of fustick, (the best kind) half a pound of ground turmerick, and one ounce of aquafortis; boil well, run your cloth, and handle till your colour pleases.

22d *TO TAKE THE COLOUR OUT OF CLOTH.*

TO twenty yards of cloth, take half a pound of oil of vitriol, put in about one quart of cold water, stir it till well mixed with the water; put it in your copper already filled, and boiling hot, with fair water; run your cloth thirty minutes, air and rince, and you may make almost any colour you please, on cloth that has had the colour taken out in this way; but you cannot if done in the way of receipt No. 20 It must be observed, that there cannot be any great quantity of cloth or goods managed in these preparations at once, without shifting the liquor; for the dye-stuff that is extracted from the cloth will overpower the preparation that dissolves the colour. I have destroyed a black of the best kind and made a good yellow, in this way.

23d. *SCARLET RED.*

TO twenty yards of cloth, take one pound of good fustick, a quarter of a pound of turmerick, six ounces of aquafortis, and half a pound of argal or red tartar, which boil till the strength is well out, (the copper being clean as possible, and the water fair) then run your cloth two hours with the dye boiling; then air, rince and shift your liquor from your copper, and fill with clean water; heat boiling hot, then take one peck of wheat bran wet with vinegar, after standing twelve hours, put it in a bag, and boil well one hour; let it drain, and squeeze it as dry as you conveniently can, run your cloth 30 minutes, air, rince and shift your liquor from your copper; clean your copper as clean as possible, fill with fair water, and heat boiling hot; then add five ounces of cochineal made fine, one ounce of red arsenick, two ounces and an half of aquafortis, two ounces of gum armoniac; boil this together till the strength is well out; then run your cloth with the dye boiling, run till your colour suits, and you will have a fine scarlet.

24*th.* *SCARLET RED.*

TO twenty yards of cloth, take one peck of wheat bran wet with vinegar, let it stand twelve hours; fill your copper with water, heat boiling hot; put the bran pudding into a bag, let it boil one hour, then run your cloth with the dye boiling forty minutes; then add a quarter of a pound of aquafortis, three quarters of a pound of argal or red tartar; run forty minutes more with the dye boiling, then air, rince and shift your liquor from your copper and fill with water; add one pound of fustick, and a quarter of a pound of turmerick, boil this one hour; then run your cloth one hour with the dye boiling, air, rince and shift the liquor from your copper; fill with water, heat boiling hot; then add six ounces of cochineal pulverized, three ounces of aquafortis, and one ounce of armoniac; let it boil well fifteen minutes; run your cloth one hour with your dye boiling, and you will have a fine scarlet.

25th. CRIMSON RED.

TO twenty yards of cloth, take three quarters of a pound of allum, three quarters of a pound of cream of tartar, and three quarters of a pound of argal; pulverize these and mix them together; fill your copper with fair water, heat boiling hot, and add this compound; stir and mix it well with the boiling water; then run your cloth one hour boiling; then air, rince and shift your liquor; fill with fair water, heat boiling hot, then take half a pound of cochineal and half a pound of cream of tartar mixed and pulverized together; then add one half of the cochineal and tartar; run your cloth three quarters of an hour with the dye boiling; then air and add of this compound by little and little, with your dye boiling, till the colour is well raised on the red; then take half a pound of the spirits of sal armoniac, and run your cloth three quarters of an hour, and this will give it the crimson hue. This is a true crimson, and permanent.

26th. FOR CRIMSON RED.

TO twenty yards of cloth; take three quarters of a pound of fustick, a quarter of a pound of turmerick, five ounces of aquafortis, fill your copper with water, add this and boil well, till the strength is well out; run your cloth one and an half hours with your dye boiling; then air, rince and shift your liquor from your copper, and wash clean : fill with fair water, heat boiling hot, then take four and an half ounces of cochineal, & four and an half ounces of cream of tartar, pulverized together ; add this to the water with a quarter of a pound of aquafortis, and three ounces of turmerick, in which boil and handle your cloth, run one hour, then take half a pound of spirits of sal armoniac, or good old sig, to bloom with ; in this handle with the dye boiling, till your colour pleases.

27th. *FOR RED WITH RED-WOOD OR NICARAGUA.*

TO twenty yards of cloth ; take ten pounds of red-wood or Nicaragua chips, and boil moderately in good clean water one hour ; then add one pound of allum, run your cloth forty minutes, then air and let the dye steep in the same manner as it did before ; and run again, adding a little allum every time you dip ; and manage in this form till your colour suits your fancy. Red-wood or Nicaragua may be mixt together or used separately, just as the dyer thinks fit and proper. I commonly use both together.

28th. *CRIMSON RED WITH RED-WOOD.*

TO twenty yards of cloth, take eight pounds of red-wood, boil well, but not fast, one hour, then add half a pound of allum, run your cloth three quarters of an hour, air and let the dye simmer in the same manner as before ; add a little allum and run your cloth, and manage in this form till the strength is well out of the dye ;

then add half a pound of pearlash and handle till your colour pleases.

The dyes for red, that are made of red-wood and Nicaragua, must not be hurried and drove, nor crowded too full, because it will destroy the lustre of the red, and the colour will be dull. It is necessary the copper and all the utensils should be clean.

29th. *FOR RED WITH MADDER.*

TO twenty yards of cloth, take one peck of wheat bran, boil it in a small kettle with eight gallons of water, one hour; then fill your copper with water, boiling hot; then add the liquor of the bran, and three and an half pounds of allum, one pound of red argal, boil and run your cloth, (being well scoured and clean) one and an half hours, boiling; then air and rince your cloth, and shift the liquor from your copper; fill with fair water, then add eight pounds of madder that is good, and heat moderately, with constant stirring, till near scalding hot; run your cloth three quarters of an hour with a moderate fire, then increase your fire, and bring it near a boiling heat, but not boiling, for the madder must not boil, if you intend to have a good red; then run your cloth in this manner until the strength is well out of the madder, and the colour well raised on the red; then shift your liquor from your copper; fill with water, and add two and an half pounds of the best Brazil, boil well one hour, and add three quarters of a pound of allum and run your cloth till your colour suits, boiling between each dipping; and this will produce a good red.

This colour may be finished in the madder

dye without shifting the dye, by adding two gallons of lant or sig. After the colour is well raised in the madder, run your cloth thirty minutes, and it will answer.

The best is with Brazil, but it is more lengthy, and the colour is brighter than with the sig; so I leave it to the discretion of the dyer.

30*th.* *FOR MERROON RED.*

TO twenty yards of cloth, take six quarts of wheat bran, wet with vinegar, let it stand twelve hours, and sour; put it in a bag, fill your copper with water, heat boiling hot, and boil the pudding two hours; then take it out and let it drain; squeeze as dry as you can conveniently; then add one and an half pounds of allum, and half a pound of red argal made fine, run your cloth one hour boiling, air and let it lie all night and sour; then rince your cloth, shift your liquor from your copper, and fill it with fair water: when warm, add ten pounds of good madder, and four quarts of wheat bran, constantly stirring until it is near boiling, but not boiling, for madder must not boil; run your cloth and manage in this manner till the strength is well out of the dye, and the red well raised, then add one gallon of lant or sig, and handle till your colour pleases.

31*st.* *FOR POLISHED RED WITH MADDER.*

TO twenty yards of cloth, take three and an half pounds of nutgalls pulverized, put them in the copper, and fill the copper about half full of water, put the galls in, let it boil till the strength is well out; then fill the copper with cold water; see that your dye it not hotter than scalding hot; then add five, six, or seven pounds of the best madder, in proportion to the shade required; let it simmer with a small fire one hour, with frequent stirring; then run your cloth thirty minutes, air and run again with the heat increasing; run till the strength is well out of the dye, and the colour well raised on the red. The dye must steep between each dipping, fifteen or twenty minutes, with the heat increasing, but not boiling, for it will destroy the substance of the madder to let it boil. If your colour is not dark enough, add a little potash or pearl-ash, and handle till your colour pleases; and you will have a fine polished red.

32d. *FOR PORTABLE RED.*

TO twenty yards of cloth, take one pound of fustick, and three quarters of a pound of allum, fill your copper with water, heat boiling hot, run your cloth, after the strength is out of the fustick, run three quarters of an hour; shift your copper, fill with fair water, and then add six pounds of red-wood, let it boil moderately one hour, then add three quarters of a pound of allum, run your cloth 40 minutes; then air, and let the dye simmer one and an half hours, and run your cloth as before; then air and take out the chips, and add one and an half ounces of cochineal, and three ounces of aquafortis; run again with the dye boiling, 40 minutes; to bloom, take six or eight ounces of spirits of sal armoniac, or good old sig; and your cloth will be a good colour by handling in this half an hour.

33d. *FOR CLARET RED.*

TO twenty yards of cloth, take two pounds of fustick chips, fill your copper with water, boil well, then add one pound of allum, boil, run your cloth one hour boiling, then air, rince and shift your copper; fill with fair water, add eight pounds of red-wood, boil well, and add half a pound of allum; run your cloth one hour, then air, let the dye steep one hour, and run again, adding a little allum; manage in this manner until the strength is well out of the dye, and the colour well raised on the red; then add two ounces of aquafortis, killed with pewter or block tin, as described in receipt 18th, run your cloth thirty minutes with the dye boiling; then add two gallons of sig to bloom, handle till your colour pleases, and you will have a fine claret red.

34th. *FOR CLARET.*

TO twenty yards of cloth, take twelve pounds of barwood, boil well, then add half a pound of allum, run your cloth until the strength is well out of the dye, about thirty minutes to a dipping, boiling between each dipping as much as is necessary to get the strength out of the barwood: when the colour is well raised on the red, then add a quarter of a pound of logwood, and a quarter of a pound of copperas mixed together, and handle until your colour pleases.

35th. *FOR MADDER RED TO BE DYED A CLARET.*

TO twenty yards of cloth, take one pound of

logwood, fill with fair water, boil well, run your cloth, and sadden with copperas until your colour pleases.

36th. *FOR SCARLET TO BE DYED CLARET OR ANY DARK COLOUR.*

TO colour twenty yards of cloth; fill your copper with water, heat boiling hot, then add one pound of copperas; run your cloth, air, and run it again; then shift your liquor from your copper, rince it, and fill with water; then add one and an half pounds of logwood, boil well twenty minutes, then run your cloth till your colour pleases; and you will have a fine claret that is durable.

This is the only way that scarlet can be coloured a darker colour. By running it in the copperas water first, you may dye it almost anydark colour you please; for the copperas will destroy all the acidous power that the scarlet is made by and depends upon; but until the power of the acid is destroyed, you cannot strike any colour through, so but that it will remain red in the middle of the cloth.

I have coloured scarlet black completely through, and almost all other dark colours, by the help of copperas.

37th. *FOR CHERRY COLOUR.*

TO twenty yards of cloth, take seven and an half pounds of barwood, boil well, and add a quarter of a pound of allum: then run your cloth one hour: air and add two pounds of Brazil, and boil till the strength is well out; run your cloth again as before till the colour is well raised on the red, then add two quarts of sig or lant, and handle till your colour pleases.

38th. FOR VIOLET COLOURS.

TO twenty yards of cloth, take four pounds of Brazil, and one and a quarter pounds of logwood; boil well, and add three quarters of a pound of allum, then run your cloth thirty minutes, air, and let it steep till the strength is well out; then run again as before, then add three quarts of lant or sig, with the dye hot and well mixed together; run your cloth, and handle till your colour pleases.

Twenty shades of violet colour may be produced, by varying the logwood and brazilletto. The further management of this dye, I have left to the fancy of the dyer, for the colour will be beautiful, almost equal to cochineal and indigo.

You may use peach-wood in part, instead of all brazilletto, if you like. It will be less expensive than all brazilletto; but this I leave to your own choice.

39th. FOR PINK COLOUR.

FOR twenty yards of cloth, fill your copper with fair water, heat boiling hot, then add two pounds of allum, and one pound of argal; in this boil and run your cloth one hour, then air, rince and shift your copper; fill with water, and add two pounds of madder. Let it heat moderately, with often stirring, till near boiling hot, run your cloth one hour; and you will have a good colour of the kind.

40th. *FOR FLESH COLOUR.*

TO twenty yards of cloth, take one and an half bushels of black birch, and half a bushel of hemlock bark, boil well till the strength is well out; then add a quarter of a pound of allum, run your cloth one hour, and handle, and you will have a good colour of the kind.

41st. *FOR ORANGE COLOUR.*

TO twenty yards of cloth, take two pounds of fustick chips, 3 ounces of argal, and half a pound of allum, boil till the strength is well out of the fustick, then run your cloth, with the dye boiling, one hour; then air, rince, and shift the liquor from your copper, and fill with fair water; then add two and three quarters pounds of red-wood, two and three quarters pounds of madder, three quarters of a pound of allum, and two ounces of aquafortis; let it boil moderately, with often stirring, till the strength is well out; then run your cloth one hour; then add one and an half ounces of arsenick, and half an ounce of cochineal, and this will bind the colour. In this run and handle till your colour pleases.

42d. *FOR ORANGE.*

TO twenty yards of cloth, take eight pounds of fustick, and four pounds of red-wood, and boil well; then add half a pound of allum, run your cloth thirty or forty minutes, then air, and let the dye steep a while, then run again till the strength is well out of the dye; then add one gallon of sig to bind; and handle till your colour suits.

43d. *FOR BROWN.*

TO twenty yards of cloth, take two bushels of butternut bark, fill with water, heat moderately, let it steep, (but not boiling) till the strength is well out of the bark; then run your cloth three quarters of an hour; and air and run again with the dye hot, but not boiling, (for boiling the bark destroys part of the lustre of the colour which the bark gives) but run in this manner till the strength is well out of the dye, then, air and take the bark out of your dye; then add a quarter of a pound of copperas and two quarts of sig, and mix the dye well together; run your cloth with your dye boiling fifteen or twenty minutes, and handle in this manner till your colour pleases.

Various shades may be produced in this dye, by varying the bark and copperas; sometimes more of one sort, and sometimes less; and thus by changing the order of them, different shades will appear. Dry bark and green will make a different shade; boiling and not boiling will have the same effect. Thus I leave it to the discretion of the dyer, to vary them as he or she pleases, to answer the shade or shades required.

FOR LONDON BROWN OR CORBEAU WITH CAMWOOD.

TO twenty yards of cloth, take five pounds

of good ground camwood, fill your copper with fair water, heat boiling hot, let your camwood boil a few minutes, then run your cloth one hour; air and run again in the same manner as before; air and add half an ounce of blue vitriol, and a quarter of pound of oil of vitriol,* boil well five or six minutes, then run your cloth twenty or thirty minutes more; then take one pound of copperas dissolved in vinegar by constant stirring on the fire, (but be sure and not let it boil, for it will spoil the dye) then add the copperas by little and little, the dye boiling, and run as before, and handle till your colour pleases. If it is not dark enough for the corbeau, take two ounces of verdigrease made fine, and dissolved in sig or vinegar on the fire, by often stirring, as described in receipt 4th; add this with one pound of logwood chips; boil well, and handle in this manner till your colour suits. Sometimes it is required to be very dark, then these darkening materials must be applied according to the judgment of the dyer, &c. You may change this colour by adding a few ounces of pearlash, to a bright purple, which will be permanent.

* When oil of vitriol is applied to any hot liquor, you must before you put it in the dye, put seven-eighths of cold water to it, and then it will heat near boiling hot with the cold water; but if you put in otherwise, it will make the hot liquor fly in a shocking manner, and the dyer will be in danger of being scalded; and another thing to be observed, you must raise your red for your body, with camwood before you apply your vitriol, or your camwood will be lost; for camwood cannot run upon any other dye stuff; in what colour it is used it must be first applied, otherwise it will be of no use; yet camwood is the best dye-wood in the world if used right.

45th. *FOR LONDON BROWN OR CORBEAU WITH NICARAGUA.*

TO twenty yards of cloth, take eight pounds of Nicaragua, and half a pound of fustick; boil well, and add half a pound of allum, run your cloth till the strength is well out of the dye, and the colour well raised on the red, then add half an ounce of blue vitriol, and half a gill of oil of vitriol, and four quarts of sig, run your cloth 30 minutes; then add half a pound of logwood, boil well, add one ounce of verdigrease, pulverized and dissolved, as in receipt No. 4, run your cloth twenty minutes; then add copperas by little and little to sadden; and handle till your colour pleases.

46th. *LONDON BROWN OR CORBEAU WITH RED-WOOD.*

TO twenty yards of cloth, take two pounds of fustick chips, boil well, and add one pound of allum, run your cloth boiling three quarters of an hour; air and rince, and shift your copper, then fill with water, and add ten pounds of red-wood chips; let it boil moderately one hour; then add half a pound of allum, run your cloth forty minutes, air, and let the dye steep one hour, and run again as before; and handle in this manner till you have a good red; (you must be cautious not to drive the dye too fast, and add a little allum now and then if necessary) and till the strength is well out of the dye: then add one gallon of sig or urine, run your cloth half an hour, then add one and an half pounds of logwood chips, boil well, then add two ounces of verdigrease made fine and dissolved in one pint of vinegar, as described before, and handle till your colour pleases.

47th. LONDON BROWN.

TO twenty yards of cloth, take two pounds of fustick and seven pounds of red wood chips, boil moderately one hour, then add half a pound of allum, run your cloth three quarters of an hour, then slacken the heat of your dye, and add three pounds of madder; let it stand and simmer with often stirring half an hour, run your cloth one hour, and if the strength is not out of the dye, run again. The cloth must be a good red before you sadden; then add copperas to sadden with by little and little, till your colour suits.

48th. FOR LONDON BROWN.

TO twenty yards of cloth, take four pounds of fustick chips, boil well, then add half a pound of allum; then run your cloth one hour boiling, then air and rince, and shift your copper and fill with fair water; then add six pounds of red-wood chips, boil well, add half a pound of allum, run your cloth one hour, then add one and an half pounds of madder, let it simmer half an hour, then run your cloth one hour, then add three quarters of a pound of logwood chips, boil well, then add two gallons of sig; then run your cloth and handle till your colour pleases.

49th. FOR REDDISH BROWN.

TO twenty yards of cloth, take one and an half

pounds of fustick, boil well, and add a quarter of a pound of allum, in which run your cloth one hour boiling; air and rince your cloth, shift your liquor from your copper and fill with fair water, then add nine pounds of red-wood; let it boil well, then add half a pound of allum, run your cloth one hour, then add a quarter of a pound of pearlash and a quarter of a pound of allum; run your cloth half an hour, and this will be a good red; then add one ounce of arsenick and a quarter of a pound of argal; run your cloth three quarters of an hour, then add two gallons of good old sig, and handle till your colour pleases, and you will have a fine colour.

50th. *FOR SPANISH BROWN.*

TO twenty yards of cloth, take one bushel of butternut bark, and one bushel of walnut bark, boil well, run your cloth one hour, then take the bark out of the dye, and add half a pound of copperas; run your cloth forty minutes; then air and rince, and shift your liquor from your copper; fill with fair water, and add two pounds of fustick chips; boil well, then add half a pound of allum, run your cloth one hour, and air and rince, and shift your liquor from your copper, fill with fair water, and add eight pounds of red-wood; boil well and add half a pound of allum, run your cloth one hour; then add two ounces of oil of vitriol, killed with the flower of brimstone; run your cloth half an hour; then add half a pound of logwood, and boil well, then add two gallons of good old sig; and handle till your colour pleases.

51st. *FOR LONDON SMOKE.*

TO twenty yards of cloth, take eight pounds of fustick chips, boil well, then add a quarter of a pound of allum; run your cloth half an hour, then add one and an half bushels of good butternut bark, boil moderately till the strength is well out, then run your cloth one hour with the dye hot; then if the strength is well out of the dye, take the bark and chips out of the dye, and add three pounds of Nicaragua wood, or red-wood, and one and an half pounds of logwood chips, boil well thirty minutes; then run your cloth one hour, then add one gallon of sig, run twenty minutes with the dye boiling, then add one and an half or two pounds of copperas, and run to your liking; and this will be a colour equal to a blue for strength, &c.

52d. *CINNAMON BROWN.*

TO twenty yards of cloth, take four pounds of fustick, and three pounds of red-wood chips, or Nicaragua, boil well, then add half a pound of allum; run your cloth one hour, then slack the heat of your dye, and add four pounds of good madder; let it simmer half an hour; then add half a pound of allum, run your cloth one hour; then add two ounces of copperas, and two gallons of sig; and handle with the dye hot till your colour pleases.

53d. *FOR SMOKE BROWN.*

TO twenty yards of cloth, take six pounds of

fustick chips, and three pounds of ground cam-wood, boil well till the strength is well out; then run your cloth one hour, then add three and an half pounds of coarse madder; let it simmer twenty minutes; then run your cloth half an hour; then add half a pound of copperas, and handle till your colour pleases.

54th FOR LIVER BROWN.

TO twenty yards of cloth, take eight pounds of fustick chips, and two pounds of red-wood chips, boil well one hour, and run your cloth forty minutes; then add four pounds of mull, or coarse madder, and two quarts of rotten wood of oak, boil moderately, and run your cloth one hour; then add six or eight ounces of copperas, and handle till your colour pleases.

55th. FOR OLIVE BROWN.

TO twenty yards of cloth, take five pounds of fustick chips, boil well, run your cloth one hour, then add one bushel of butternut bark; boil well, but moderately, one hour; then run your cloth one hour, or till the strength is well out of the dye; then take the bark and chips out of the dye, and add six ounces of copperas, and handle till your colour pleases.

56th. FOR OLIVE BROWN.

TO twenty yards of cloth, take six pounds of fustick chips, and one pound of logwood, boil well, and run your cloth half an hour; then add one pound of madder, let it simmer half an hour, then run your cloth as before; then add a quarter of a pound of chymick or blueing, stir and mix it well with the dye, and run your cloth twenty minutes; then add one and an half pounds of logwood, and one gallon of sig; run your cloth as before, add six ounces of copperas, and handle till your colour pleases.

57th. *FOR OLIVE BROWN.*

TO twenty yards of cloth, take seven pounds of fustick chips, three quarters of a pound of logwood, and half a pound of madder; boil well one hour, then run your cloth one hour, then add half a pound of chymick or blueing, and run your cloth twenty minutes; then add two quarts of sig, and run again as before; then add two ounces of copperas, and handle till your colour pleases.

58th. *FOR A LIGHT SNUFF BROWN.*

TO twenty yards of cloth, take eight pounds of fustick chips, and four pounds of red-wood or Nicaragua; boil well an hour and a half, then add a quarter of a pound of allum; run your cloth thirty minutes, then air and run again till the strength is well out of the dye; then add one gallon of sig, run your cloth half an hour, then take one peck of soot scraped from the chimney, put it into a tub, and put two pails full of your dye to it; stir it well together, and let it stand and settle; then pour off the liquor moderately, and add it to your dye; run your cloth, and handle till your colour suits.

59th. *FOR SNUFF BROWN.*

TO twenty yards of cloth, take four pounds of fustick chips, and boil well; then add a quarter of a pound of allum, and run your cloth half an hour; add five pounds of red-wood, boil well, and then add half a pound of allum; run your cloth as before till the strength is well out of your dye, then add a quarter of a pound of argal, and handle till your colour pleases.

60th. *FOR DARK SNUFF BROWN.*

TO twenty yards of cloth, take six pounds of fustick chips, and boil well, then add a quarter of a pound of allum; run your cloth one hour, then add two pounds of ground camwood, and one and an half pounds of madder, and let it simmer half an hour; run your cloth one hour, then add half a pound of copperas, or more, if the colour is not dark enough; and handle till your colour pleases.

61st. *FOR SNUFF BROWN.*

TO twenty yards of cloth, take three quarters of a bushel of butternut bark, and three quarters of a bushel of walnut bark, boil well one hour, but moderately; run your cloth one hour, then if the strength is well out of the bark and dye, take the bark out of the dye, and add one pound of copperas to sadden with; run your cloth three quarters of an hour, air and rince your cloth and shift your liquor from your copper, wash clean and fill with fair water; then add four pounds of fustick chips, boil well, and then add half a pound of allum: run your cloth half an hour; then add five pounds of red-wood chips, boil one hour, and add a quarter of a pound of allum; run your cloth three quarters of an hour; let it steep, and run till the strength is well out of the dye. To sadden, take one gallon of sig, and handle, &c.

62*d*. *FOR SNUFF BROWN.*

TO twenty yards of cloth, take one pound of allum, boil, and run your cloth one hour, then shift your liquor from your copper, and fill with fair water; then add five pounds of fustick, boil well till the strength is well out, then run your cloth thirty minutes; then add one bushel of butternut bark, and five pounds of sumac berries, boil moderately one hour, and then run your cloth forty minutes; then add six ounces of aquafortis, killed with pewter, as described before in receipt No. 18; run your cloth with the dye boiling one hour, and the colour will be done.

63*d*. *FOR SNUFF BROWN.*

TO twenty yards of cloth, take eight pounds of fustick chips, boil well, and add a quarter of a pound of allum; run your cloth thirty minutes, then add four pounds of red-wood chips or two pounds of ground camwood ; boil well, and run your cloth till the strength is well out of the dye ; then add one gallon of sig, a quarter of a pound of logwood, and an ounce of verdigrease, prepared as in receipt 4th ; boil well, run your cloth twenty minutes, then add two ounces of copperas, and handle till your colour pleases.

64th. *FOR SNUFF BROWN.*

TO twenty yards of cloth, take eight and an half pounds of fustick chips, four pounds of coarse madder, and three quarters of a pound of logwood ; boil well till the strength is well out of the dye-wood, but not fast ; or the madder may be omitted till the strength is boiled out of the logwood and fustick, and then let it simmer a short time ; then add six ounces of allum, run your cloth one hour, air, and run again, till the strength is well out of the dye ; then add half a pound of copperas to sadden, or more if it is not dark enough; and handle till your colour pleases.

65th. *FOR BAT-WING BROWN.*

TO twenty yards of cloth, take one and an half pounds of fustick, and four pounds of good logwood, boil well, and then add one and an half pounds of good madder, and six ounces of allum ; let it simmer half an hour, then run your cloth one hour ; add eight or ten ounces of copperas, and one quart of lant, then run and handle till your colour pleases.

If you wish to alter the shade of this colour, you may add five or six pounds of logwood, and less fustick, and you may have the colour to suit your fancy.

66th. *FOR SLATE BROWN.*

TO twenty yards of cloth, take one bushel of butternut bark, boil well and run your cloth one hour; then take out the bark, and add half a pound of copperas; run twenty minutes, air, and run again, and add more copperas if it is not dark enough; for it requires to be very dark. When dark enough, shift your copper, scour clean, and rince your cloth; fill with fair water, heat hot, then add three ounces of compound or blueing; run your cloth twenty minntes, air, and if your colour is not blue enough, add a little more blueing; and if it is not dark enough, and the colour grows lighter, then add four or six ounces of logwood, and one ounce of blue vitriol; and handle till it suits your fancy.

67th. *FOR DOVE OR LEAD BROWN.*

TO twenty yards of cloth, take half a peck of chesnut or maple bark, and two ounces of logwood, boil well, then add two ounces of copperas, and a little compound or blueing, (say half an ounce) and stir your dye well together; run your cloth twenty minutes; then if you find your colour wants altering, it may be done by varying thus;—If it is not dark enough, add a little more copperas—if not blue enough, add a little more blueing—if not bright enough, add a little more logwood; run again, and if it requires nothing, your colour will be finished. Silk may be dyed in this.

68*th*. *FOR PEARL OR SILVER GREY.*

TO twenty yards of cloth, take four quarts of wheat bran, put it in a bag, and fill your copper with fair water, and boil the pudding an hour and a half; then take it out, let it drain, and squeeze it as dry as you can; then add two ounces of allum, let it boil, and skim off the scum that will rise, then run your cloth one hour; add four pounds of logwood chips, put them in a bag, and boil well till the strength is well out, then take the bag of logwood out of the dye, if you do not, it will spot the cloth; run your cloth thirty minutes, then add half an ounce of blue vitriol, and handle till your colour pleases.

It requires care with this colour, as well as all other light colours, that you do not let the cloth touch any thing that will spot it, for there is not much, if any, remedy for a light colour when spotted; and all light colours should be dried with the backside to the sun; for the sun is apt to injure the colour.

69*th*. *FOR LIGHT BROWN.*

TO twenty yards of cloth, take half a peck of hemlock bark, with the moss taken off, and two ounces of logwood chips, boil well, run your cloth twenty minutes, then add two ounces of copperas, and handle till your colour pleases.

70th. FOR ASH BROWN.

TO twenty yards of cloth, take three quarts of white ash bark, three ounces of logwood chips, boil well, run your cloth twenty minutes : then add three ounces of copperas, and handle till your colour pleases.

71st. FOR DRAB BROWN.

TO twenty yards of cloth, take a half peck of chesnut or maple bark, green or dry, two pounds of fustick chips, and two ounces of logwood chips : boil well, then add one ounce of compound of blueing, run your cloth twenty minutes : then add two ounces of copperas, and handle till your colour pleases.

72d. FOR DRAB.

TAKE chesnut, black birch, and yellow oak bark, half a peck of each, boil well, run your cloth, then add three ounces of copperas ; and handle till your colour pleases.

73d. FOR DRAB.

TAKE one quarter of a pound of nutgalls, made fine, then one quarter of a pound of fustick, boil well, run your cloth ; then add half an ounce of blue vitriol, two ounces of copperas ; run your cloth fifteen minutes, then add half a gill of oil of vitriol and one ounce of blueing, and stir it well with the dye, run your cloth, and handle till your colour suits.

74th. FOR DRAB.

TAKE six ounces of nutgalls, pulverized, three ounces of the flour of brimstone, four ounces of allum—put them in fair water, run your cloth one hour; then sadden with black float, and handle till your colour suits.

75th. FOR DRAB.

TAKE one and an half pounds of fustick, one pound of logwood, one quart of rotten wood of oak, boil well, then add one half pound of madder, and four ounces of allum, boil, run your cloth twenty minutes; then add three ounces of copperas and one quart of sig, and handle till your colour pleases.

76th. FOR DRAB.

TAKE one and an half pounds of fustick chips, six ounces of logwood, boil well; then add one quarter of a pound of allum, run your cloth thirty minutes; then add three ounces of copperas, and handle till your colour pleases.

77th. FOR FOREST CLOTH.

TAKE two pounds of fustick chips, six

ounces of logwood, boil well, then add seven ounces of chymick, run your cloth twenty minutes; then add three ounces of good madder, two ounces of red tartar, made fine—let it simmer fifteen minutes, and run your cloth twenty minutes: then add one gallon of sig, or lant, and three ounces of copperas, and handle till your colour pleases.

78th. FOR LIVER DRAB.

TAKE one pound of fustick chips, three pounds of rotten wood of oak, three ounces of barwood, two ounces of logwood chips, one pound of madder, boil well, run your cloth twenty minutes; then add six ounces of filings of iron, boil well, run your cloth fifteen minutes: then add six ounces of logwood, and five ounces of copperas, and handle till your colour pleases.

79th. FOR LIGHT LIVER DRAB.

TAKE two ounces of blue galls, one ounce of logwood, two ounces of allum, one ounce of cream of tartar, and two ounces of madder: run your cloth fifteen minutes, then add one ounce of copperas, and handle till your colour pleases.

80th. FOR A MADDER DRAB.

TAKE three pounds of good madder, one pound of fustick, let it simmer one hour; then add two ounces of allum, run your cloth half an hour; then add one pound six ounces of filings of iron, boil well, run your cloth; then add three ounces of logwood, and handle till your colour pleases.

81st. *FOR A GREEN DRAB.*

TAKE three quarters of a pound of fustick, one quarter of a pound of logwood chips, boil well, then add half a pound of allum, two ounces of blueing: mix it well with the dye, run your cloth, thirty minutes; then add one ounce of copperas, and handle till your colour suits your fancy.

82d. *FOR A REDDISH DRAB.*

TAKE three ounces of allum, half a pound of fustick, six ounces of logwood chips, two ounces of madder, add two ounces of camwood, one and an half pints of rotten wood of oak; boil well half an hour, run your cloth one hour, air, sadden with three ounces of copperas: and handle, till your colour pleases.

83d. *FOR REDDISH DRAB.*

TAKE one and an half pounds of fustick, boil well; then add one quarter of a pound of allum, run your cloth boiling, one hour, then air and rince and shift the liquor from your copper, fill with fair water; then add three and an half pounds of good madder, two ounces of camwood, let it simmer, fifteen minutes; then run your cloth twenty minutes, then add two ounces of filings of iron, and handle till your colour pleases.

84th. *FOR LIGHT DRAB.*

TAKE five ounces of fustick chips, two ounces of good madder, two ounces of allum, boil well, run your cloth twenty minutes; then sadden with two ounces of copperas, and handle till your colour pleases.

85th. *FOR YELLOW DRAB.*

TAKE three quarters of a pound of fustick, two ounces of madder, two ounces of logwood, boil well; then add one quarter of a pound of allum, run your cloth one hour; then sadden with two ounces of copperas, and handle till your colour pleases.

86th. *FOR A DARK YELLOW DRAB.*

TAKE two pound of fustick chips, five ounces of logwood chips, boil well, then add five ounces of madder and one quarter of a pound of allum, run your cloth thirty minutes, then add one quarter of a pound of copperas, and handle till your colour pleases.

87th. *FOR A FOREST BROWN.*

TAKE six pounds of fustick chips, boil well:

then add two ounces of allum, run your cloth fifteen minutes; then add two and an half pounds of logwood, boil well, run your cloth thirty minutes, then sadden till your colour suits, with six ounces of copperas.

88th. *FOR A DARK FOREST BROWN.*

TAKE one and an half pounds of logwood, three quarters of a pound of red argal, and three quarters of a pound of allum, boil well, run your cloth one hour, boiling; then add four pounds of good fustick chips, boil well, run your cloth half an hour, and handle till your colour pleases.

89th. *FOR PARIS MUD*

TAKE your cloth, and dye it a bright lively blue, but not deep; then rince your cloth, and fill your copper with fair water; then add six pounds of stone rag, or the moss of stone, boil well, run your cloth one hour; then add two ounces of copperas, and one quart of sig, and handle till your colour pleases.

90th. *FOR A RAVEN COLOUR.*

TO twenty yards of cloth, take two quarts of wheat bran, wet with vinegar; let it stand two days and sour, then fill your copper with fair water, put the bran into a bag, boil well one hour; then take out the bag and let it drain, then add one pound of madder and one pound of allum; run your cloth one and an half hours, boiling: then air and fold it up smooth, and wrap it up close, and let it lie twenty-four hours; then rince, and shift the liquor from your copper, fill with fair water, then add eight pounds of logwood chips, boil well till the strength is well out; then run your cloth one hour; then, if you find it necessary, add more logwood—if not, then add one quarter of a pound of copperas, and one gallon of lant, and handle till your colour pleases.

If your colour is not dark enough, you may use a little ashes, put with sig; and take the lie and put in the dye, with a little copperas, and run again.—Lie and sig has the same effect, and potash or pearlash.

91st. *FOR CROW, WITH COPPERAS.*

TO twenty yards of cloth, take one and an half pounds of copperas, fill your copper with water, heat boiling hot; then run your cloth twenty minutes, air, and run again as before, then air and rince your cloth, shift the liquor from your copper, and rince, fill with fair water, heat, and add four pounds of logwood chips, boil well, run your cloth half an hour, then air and run again as before; then, if your colour is not dark enough, add one ounce of blue vitriol, run again, and handle till your colour pleases.

92d. *FOR CROW, WITH BLUEING COMPOUND.*

TO twenty yards of cloth—fill your copper

with fair water, heat boiling hot, then add one pound of blueing, (made as in receipt No. 6, for Prussian blue) add this at twice or three times, run your cloth twenty minutes at a time, air and stir the blueing well with the dye, before the cloth is dipped in the dye; then add two pounds of logwood chips, boil well, then add one quarter of a pound of verdigrease pulverized and dissolved in vinegar, as in receipt No. 4; then run your cloth half an hour, then add half a pound of copperas, run again, air, and if it is not dark enough, add more copperas, and handle till your colour suits your fancy.

93*d*. *FOR CROW, WITH BLUE VITRIOL.*

TO twenty yards of cloth —Fill your copper with water, heat scalding hot, take half a pound of blue vitriol, let it dissolve, run your cloth forty minutes, in two parts: then add five pounds of logwood chips, boil well, run your cloth thirty minutes, air and run again, and handle till your colour pleases.

94*th*. *FOR BLACK.*

TO twenty yards of cloth—Fill your copper with water, heat, and add three pounds of copperas; heat near boiling, run your cloth one hour, then air and run again, boiling the time as before: air and rince, and shift the liquor from your copper (rince your copper clean) and fill with water, and add six pounds of logwood chips, boil well, run your cloth thirty or forty minutes, let it boil again fifteen or twenty minutes, then run again as before; then add one quarter of a pound of blue vitriol, run your cloth, boiling, three quarters of an hour ; then, if it is not black enough, run again, and handle till your colour pleases.

This is the best form to dye a black, I think, in the world ; it is equal to any for brightness, and without the least danger of rotting the cloth ; and the colour is lasting and permanent as a blue or scarlet.

It is necessary to cleanse the colour or dye stuff well out of the cloth, immediately. First rince in fair water, then take a tub of warm water, sufficient to handle, and wet the before-mentioned quantity of cloth ; then add half a pint of the liquor of beef galls, mix it well with the warm water, then handle your cloth in this till it is well wet, then rince in water till it is clean. This is a sure remedy against crocking. The beef gall may be used in all cloths, in this manner, that are liable to crock ; and it will prevent their crocking, without the least danger of injuring the colour.

95th. FOR BLACK.

TO twenty yards of cloth, take three pounds of logwood chips, one and an half pounds of sumac, of one season's growth, cut and dried : boil well, run your cloth half an hour, then add one ounce of blue vitriol, one quarter of a pound of nutgalls, pulverized, boil well, run your cloth fifteen minutes : then add one ounce of verdigrease, pulverized and dissolved in sig or vinegar, as described in receipt No. 4 : run your cloth fifteen minutes, then add one pound of copperas, handle, and if it is not black, then add more copperas ; and handle till your colour pleases.

96th. FOR BLACK.

TO twenty yards of cloth, take six pounds of logwood chips, one pound of dry alder bark one and an half pounds of sumac, of one, season's growth, well cured and dried, one quarter of a pound of fustick, boil well one hour, then run your cloth one hour, air and run again as before; then air, add one gallon of sig, and one and an half pounds of copperas, run your cloth twenty minutes; then if it is not black, add more copperas, and if it is attended with a rusty brown, add two pounds of common good brown ashes, run your cloth, and handle till the strength is well out of the dye.

Then, if it is not black, shift your liquor from your copper, scour clean, rince your cloth, fill your copper with fair water, then add one pound of logwood chips, one quarter of a pound of alder bark and half a pound of argal; then boil well, run your cloth one hour, then sadden with copperas, what is necessary, and handle. But if it continues of a rusty cast, which logwood causes, add one gallon of sig, or more ashes, that which is most convenient, and handle till your colour pleases.

N. B. Silk may be dyed in this dye. It is necessary to take the same method in cleansing as in receipt No. 94, and all other dark colours that are liable to crock, &c.

97th. FOR BLACK.

TO twenty yards of cloth, take three quarters of a pound of blue vitriol, add to fair water, boil well, run your cloth three quarters of an hour; then add six pounds of logwood chips, and one pound of fustick chips, boil one hour, run your cloth one hour, then add two ounces of verdigrease, pulverized and dissolved in vinegar, as before described, and one gallon of sig, run your cloth twenty minutes; then add one pound of copperas, and handle with the dye boiling, till your colour pleases.

98th. *FOR BLACK.*

TO twenty yards of cloth, take one bushel of butternut or chesnut bark, or both mixed together: boil till the strength is well out, then run your cloth one hour, then sadden with copperas till it is quite dark; then air and rince, and shift your copper, fill with fair water; then add four pounds of logwood chips, half a pound of fustick chips, boil well till the strength is well out, then run your cloth one hour; air, and if it is not black, or near a black, run again; then add one pound of copperas, and one gallon of sig; boil well, run your cloth boiling, and handle till your colour suits your fancy.

The preceding Receipts are calculated for twenty yards of fulled cloth; but thin cloth may be dyed as well as thick, and all kinds of woollen goods, as yarn, wool, &c. Silks may be dyed in most of the dyes before mentioned; but the dye requires to be stronger for silk than for woollen. Those dyes that will not answer for silk, I shall mention hereafter.

RECEIPTS

FOR COTTON AND LINEN,

COLD AND HOT.

99th. *BLUE*—FOR *COTTON*, *LINEN*, *YARN*, *&c.*

TO a tub that will hold thirty-six pails of water, take twelve pounds of stone-lime, slack it, put it in, stir it ten or twelve minutes; then add six pounds of copperas, dissolved with hot water, stir it as before; then add six pounds of indigo, ground fine, stir it incessantly two hours; for three days, stir it three or four times in a day, then let it stand fifteen or twenty hours before the yarn is put in, lay sticks across the tub, to hang the yarn on, that it may not reach the bottom; move the yarn round every fifteen minutes. Six hours is sufficient for the first colouring of the dye; as the dye grows weaker, longer time is required: rince and dry it in the shade.

When the dye is reduced, then recruit in manner and form as in setting, only when there is a great quantity of sediment at the bottom, then the dye must be dipped off, leaving the sediment in the bottom; then throw away the sediment, shift the dye back, and if the tub is not full enough, then add more water, (rain water is required in this dye in setting and recruiting). The dye must not be worked at too soon after recruiting, or sitting, and it must not be crowded too full in colouring, but judgment must be used by the dyer, &c.

100th. *BLUE—FOR COTTON AND LINEN COLD.*

TO set a tub of twelve gallons, take ten gallons of good sig, to which add three gills of spirits, one pound of good indigo, three ounces of pearlash, a quarter of a pound of good madder, and a pint of wheat bran; put the indigo in a bag, and rub it in the dye till the indigo is dissolved, and stir the dye well together with the ingredients; let it stand twelve hours covered close and kept warm, and manage it in the manner and form as in receipt No. 2, till the dye comes to work. After the dye has come to work, wet the yarn in hot water, with a little pearlash in it; let it cool, then put it in the dye loose; let it lie in the dye twelve hours, then wring it out and let it air; and if it is not dark enough, then put it in again. There ought to be something at the bottom to keep the yarn off of the sediment.

There may be a saving in colouring cotton or linen, by first colouring brown or purple, as I shall hereafter mention. Silk may be dyed in this dye, but not in the blue vat.

101st. *BLUE—FOR COTTON AND LINEN—HOT.*

HEAT water sufficient for your yarn, say for five pounds of cotton or linen yarn, take five ounces of blue vitriol, run your yarn or let it lie in the dye one hour, then add three pounds of good logwood chips, boil well, and put in the yarn; let it lie one hour, then air and add two ounces of pearlash, let it lie thirty minutes; then, if it is not dark enough, add a little blue vitriol; put it in again, and you will have a good looking blue, but it will not be so lasting a colour as the two forms before mentioned.

102*d.* *To take the Colour out of Silk, Cotton, or Linen, when spotted or another colour is wished.—Hot.*

TO one barrel of hot water, take half a gill of oil of vitriol, put in the goods; run them fifteen minutes, air and rince them in fair water immediately, lest it should endanger the goods.

I have reduced black without injuring it, and made a yellow of it in this form.

103*d.* *FOR GREEN ON SILK—HOT.*

TAKE two pounds of fustick, boil well, till the strength is well out, then take out the chips, and add a quarter of a pound of allum, and six ounces of blueing, prepared as in receipt No. 6; stir it with the dye till it is well mixed, then handle your silk fifteen or twenty minutes: stir it lively, and keep it open and loose in the dye; (silk should never be wenched as woollen goods) air, and if not deep enough, add a little more blueing; and if not yellow enough, then a little allum, run again fifteen minutes; then air, and if the colour suits, rince immediately. The dye ought to be so fixed as to colour quick, and there must not be a great quantity coloured at once in a dye: for the dye will get too strong with the vitriol, which will endanger the silk; but with proper care it may be coloured without any danger.

104th. *GREEN ON COTTON OR LINEN.—HOT.*

TO set a dye, take two pounds of logwood, and one pound of fustick chips, boil well, then add a quarter of a pound of allum, and run your goods one hour; then add a quarter of a pound of blue vitriol, run your goods thirty minutes, then add two ounces of pearlash; run again, and handle till your colour pleases.

105th. *YELLOW ON COTTON AND LINEN.—HOT.*

TAKE two pounds of the leaves or peelings of onions that are clean and clear from dirt; put them in fair water, boil well, then add half a pound of allum, run your goods one hour, and you will have a good colour.

106th. *ORANGE COLOUR ON COTTON AND LINEN.*

TAKE two pounds of copperas, dissolve it in hot water, and have the liquor very strong; let it stand till nearly cold, run your goods one hour, then dip it in good lye, handle till perfectly wet; then let it drain, and hang it in the sun fifteen minutes, and the sun will turn the colour; continue to manage in this manner, dipping it in the dye and hanging it in the sun, till dark enough.

107th. *FLESH-COLOUR ON COTTON AND LINEN.—HOT.*

TAKE one and an half bushels of black-birch

bark, and half a bushel of hemlock bark boil well; then add a quarter of a pound of allum, and two ounces of pearlash; run your cloth or goods till your colour pleases.

108*th*. *RED ON COTTON OR LINEN.—COLD.*

TAKE six pounds of Nicaragua chips, boil them till the strength is well out; then add half a pound of allum, and let it stand till cold; run your cloth or yarn in hot water, with a little pearlash in it; then air, and put it in the dye, frequently handling over till the colour suits.

109*th*. *COTTON AND LINEN REDISH BROWN. HOT.*

TAKE butternut, sassafras, black alder, and hemlock bark, a bushel of each; boil well, run your goods one hour, then add two pailfulls of lie, or a quarter of a pound of pearlash; run your cloth or goods, and handle till your colour pleases.

110*th*. *FOR PLUMB-COLOUR OR PURPLE, ON SILKS.—HOT.*

TAKE six pounds of logwood chips, and three pounds of redwood chips, boil well till the strength is well out of the chips; then add one pound of allum, and run your goods one hour; then add one ounce of verdigrease, made fine and dissolved in sig, described before, and add one gallon of sig; run your goods thirty or forty minutes, and if your colour is not dark enough, then add a little blue vitriol, and handle till your colour pleases.

111th. *PURPLE ON COTTON OR LINEN.—COLD.*

TAKE three pounds of logwood chips, boil well, till the strength is well out and the dye very strong, (for all cotton dyes require to be strong;) then add half a pound of allum, and one ounce of pearlash; let it stand and get cold, dip your goods into hot water, air, and put them into the dye loose, handle over once in fifteen or twenty minutes; let them lie in the dye in this manner till the colour suits. It must be observed in dying cottons and linens in cold dyes, that the air and sun are very necessary to brighten and strike the colour in. Let the goods lie in the air and sun, three or four times in the course of your colouring, fifteen or twenty minutes at a time. The preparation is suitable for blue, as mentioned in receipt 99th.

112th. *BROWN ON COTTON AND LINEN.—COLD.*

TAKE of maple or white oak bark, one bushel, boil well till the strength is well out, then take the bark out, and have dye sufficient to wet the goods; then add one pound of copperas, let it stand till nearly cold; run your goods in hot water with a little pearlash first: then put it in the dye, and handle over once in ten or fifteen minutes, and air, as described before in receipt 110th; and handle in this manner till the colour suits; then rince clean. This is the brown mentioned in receipt 99th, for a saving in blue; but I prefer the purple; but when coloured blue, after it is dry, it is necessary to scald it in salt and water, to bind the colour.

113*th*. *DOVE ON LEAD-COLOUR, ON COTTON OR LINEN—COLD.*

TAKE one pound of nutgalls pulverized, boil in water one hour, then add two pounds of copperas; let it stand till cold, and have liquor enough to wet the goods; (it requires to be very strong) put your goods in the liquor, and handle once in five or six minutes, wring and air once in half an hour; dip in this manner three hours, then rince. This liquor ought to be put in a tub, and another liquor prepared in another tub, in this manner, viz.—take six pounds of sumac, of one year's growth, cut and well dried with the leaves all on, in the summer season, and three pounds of logwood chips, boil well till the strength is well out, then shift it in the tub, and let it stand till cold; then run your goods in the same manner as before described, handle in this two hours; if the colour is not then dark enough, run again in the copperas and galls liquor, then rince and run in the logwood again, and handle in this manner till your colour suits.

N. B. Cotton and linen, when dyed in cold dyes, must always be wet and run in hot water half an hour, and then aired; and a little pearl-ash is good in the water, to cleanse the goods for colouring, &c.

Cold dyes will remain good always if properly recruited.

114th. *OLIVE ON COTTON AND LINEN. COLD.*

TAKE one pound of nutgalls pulverized, put them in water, boil one hour, then put it in a tub, then add two pounds of copperas, have the liquor strong, and enough of it to wet and cover the goods; then dip in the hot water; then stir the galls and copperas together, then put in your goods and handle over once in five minutes, that no part shall be confined, wring and air every half hour; handle in this liquor two hours, then rince, then add three pounds of fustick and one pound of logwood chips, boil well till the strength is well out; then add five ounces of good madder, and two ounces of allum; let it simmer a few minutes, then shift the liquor into a tub, and let it stand till cold; then handl your goods in the first liquor two or three houis till the colour is well raised; and if it is not dark enough, then take two pounds of fustick, and one pound of logwood, boil well; let it cool, and sadden with copperas as much as is necessary, and handle till your colour pleases.

115th. *OLIVE ON SILK, COTTON, OR LINEN.—HOT.*

TAKE five pounds of fustick, and two pounds of logwood chips, boil well; then add a quarter of a pound of blue vitriol, and a quarter of a pound of allum, run your goods one hour; then add one pound of copperas, and handle till your colour pleases. If the colour is not dark enough, you may add more copperas, &c.

116th. *LIGHT OLIVE ON COTTON AND LINEN.—HOT.*

TAKE four pounds of fustick chips, and half a pound of logwood chips, boil well, then add two ounces of allum, and one ounce of blue vitriol; then run your goods till the strength is well out of the dye; then sadden with copperas to your liking, and handle till your colour pleases.

117th. *SLATE COLOUR ON COTTON AND LINEN.—HOT.*

TAKE hot water, and dissolve one pound of copperas; run your goods forty minutes, then air and rince, and shift your liquor from your copper; fill with fair water; then add three pounds of logwood, boil well, run your goods one hour, then add a quarter of a pound of blue vitriol, and handle till your colour pleases.

118th. *BLACK ON COTTON AND LINEN.—HOT.*

TAKE four pounds of good logwood, and two pounds of fustick chips, boil well; then add a quarter of a pound of blue vitriol, run your cloth one hour, or till the strength is well out of the dye, then sadden with two pounds of copperas, and one gallon of good old sig; run your cloth, and if it is not black, you must air and rince, and shift your liquor from your copper, and set another dye in manner and form as the first, and handle again, and depend on having an excellent black at last. But if it is attended with a rusty brownness, you may put in one quart of brown ashes, or two ounces of pearlash, and handle lively, which is necessary in all hot silk, cotton, and linen dyes.

119*th*. *BLACK ON COTTON AND LINEN—COLD.*

TAKE one pound of nutgalls pulverized, boil in one pail full of water one hour, then add two pounds of copperas, shift it into a tub, and add water sufficient to cover, and handle your goods very strong; then take fair water and fill your copper, add four pounds of logwood chips, two pounds of sumac well dried, of one season's growth, and one pound of dry alder bark, boil well till the strength is well out, then dip off the dye into a tub, the chips remaining in the kettle; let it stand till cold.

The dye must be managed in this manner;—first run your goods in hot water, with a little pearlash in it; run in this half an hour, then air and lay your goods into the copperas and galls liquor; handle over every eight or ten minutes, and air every half hour; handle in this two hours, then rince clean and lay it in the logwood liquor; handle as in the other three hours, then if it is not black, put water in the copper upon the chips; before running in the copper, let it steep and cool again, and add one pound of copperas; run in this one hour; but if it has a rusty brown appearance, which is occasioned by the logwood, then add two ounces of pearlash, or brown ashes will answer if you have no pearlash; run in this half an hour, then air and rince clean, and if it is not black, then recruit the liquors and make them stronger, and manage as before in the first preparation; and never fear but you will have a fine black.

After you have rinced clean, to keep it from crocking, use beef galls, as mentioned in receipt No. 94.

GENERAL OBSERVATIONS.

COTTON and linen dye is the best cold in general; for it is almost impossible with me to colour cotton and linen in hot dyes without spotting; for the cotton, &c. are of a cold deadly nature, and the steam of the dye has a bad effect on goods of this kind. All kinds of cotton and linen cloths, yarn and thread, may be coloured by following the preceding receipts for dying cotton and linen.

In the receipts for dying silk, cotton and linen, I have not specified any particular quantity of yards or weight. There is so much difference in the weight of goods of this kind, that no rule could be given in yards; and no certainty can be affixed to a general rule of weight, because of the difference of the quality of the goods. Silks differ, so do cottons and linens; no regular system can therefore be adopted. The dyer is to proportion his dyes according to the receipts, following his judgment as the goods vary; and if he closely pursues the directions for proportion and management, he will not find a single receipt that will not answer the purpose designed. I shall hereafter speak particularly of the powers on which the dyes depend.

DIRECTIONS FOR DRESSING CLOTH.

IN dressing cloth, there are various forms in use with almost every workman in the business; but I shall only point out the way which I conceive to be the best. There are also different kinds of tools and utensils made use of, which I shall leave to the discretion of the practitioner.

FOR FULLING CLOTH.

CLOTH to be fulled, should be wet with soap sufficient to cleanse it of the dirt and grease, then scoured clean and dried; then burl or pick out all the knots and specks that will injure the cloth in dressing; then wet with soap so that the cloth will work and turn lively in the mill. Let proper attention be paid to handle the cloths from the mill, so as to keep them smooth; and be cautious not to let them grow together, for it is very hurtful to the cloth, and detrimental in dressing. The fulling-mill must be tended with care. When the cloths are fulled sufficiently, then scour clean from the soap: And if there is any of the first quality to be dressed, then card lightly over, so as to lay and straighten the nap; then shear this nap off; then take clothier's jacks, and raise a nap sufficient to cover the thread; then shear this off and raise another nap with teazles. I prefer teazles to any thing else to raise a nap on cloth; they are much milder and softer to cloth than jacks; but where they cannot be had, jacks may be substituted in their place. After raising the third nap, then colour the cloth; cleanse it well from the dye, and lay the nap straight and smooth out of warm water with jacks that are limber;

then dry, keeping the nap smooth: when dry, first shear on the back-side, then shear smooth and even on the face side, and as close as you can. When sheared, burl clean, and lay the nap with a sand-board or brick, or brush, but not with a jack; some erroneously use a jack; a jack is good and necessary to raise a nap, but not to lay it. Lay the nap smooth with the sand-board, and then the cloth is fit for the press. Have smooth papers, put it in the press, let the heat of the plate be just hissing hot; screw it moderately in the press, for the beauty of most thick cloths is destroyed by pressing too hard. The beauty of thick cloth depends on dressing and not on pressing; the coarser the cloth is, the harder it requires to be screwed; all thick cloths are not dressed alike, but according to quality, some requiring once shearing, some twice, and so on, to the number of times mentioned before; six times is sufficient for the first quality, managed as before mentioned. Some fulled cloths do not require shearing, which are dressed with a thick nap, sufficient to cover the thread; this may be raised with common wool and cotton cards; this kind of cloth is called bear-skin or coating. Bear-skin should be pressed in the cold press, never in the hot-press. Baize or flannels should be fulled lightly, the grease and dirt scoured out clean; then, if it is to be coloured, dye and raise a nap with a mild easy card or jack and a stuffed board, and dry smooth, and press in a cold press; but if it is to remain white, raise a nap as before, and dry smooth; then have a stove, or some proper tight place, with conveniences to hang the cloth up loose; then, to 100 yards of flannel, burn one pound of sulphur or brimstone under the cloths, and it will cleanse them from all pecks of dirt, and leave them as white as need be; but when you find it necessaty, you may have your copper cleaned with fair hot water, with a little compound of blueing in it; run your cloth in this a few minutes, and dry smooth; put in clean papers, press in the cold press, &c. Some, when they stove their cloth with sulphur, wet it in clean soap suds, and hang the cloth or goods up wet; but I prefer the water with a little blueing, to whiten the cloth before stoving, for it will wear handsomer, and will not grow yellow so soon.

FOR THIN CLOTHS.

THIN cloths should be well coloured, cleansed well from the dye, dried smooth, and pressed double; thin cloths require to be much moister than thick cloths; the press papers should be hard, thin and smooth; and the press hotter than for thick cloths. It must be screwed very hard, for the beauty of thin cloth is in the gloss given by pressing. The heat of the press should be kept regular, and the cloth will be smooth &c.

TO DRESS SILK AND COTTON, &c.

SILK must never be pressed, but cleaned well from the dye-stuff, then dried; then dissolve gum Arabic in water, wet the silk thoroughly in this, wring and squeeze as dry as you can, so as it shall not drip: then strain it out smooth every way, and dry. This will finish the silk dressing.

Cottons. Some do not require to be pressed, as velvets, corduroys, and similar cloths; they require only to have the nap laid when wet; fustian must have a nap raised dry with teazles, and then pressed. Almost all kinds of cotton and linen cloths, except those before mentioned, such as nankeens, jeans, muslins, &c. require to be pressed quite hard; not as hard as thin woollen cloths, but harder than thick. If any of the goods requires to be glazed, it must be managed in a different form, instead of pressing it must be calindered; *i. e.* run through a machine with two steel rollers, one hot and the other cold, and the goods rolled between them, &c.

N. B. Siik, cotton and linen, cloth must never be put in the fulling-mill to scour at any time, for it will ruin them.

Observations on the difference of Colours, and their depending powers, with directions as to the use of Dye-stuffs, and their properties and effects.

THE five Material Colours are these, Blue, Yellow, Red, Brown and Black; the three powers are these, the Alkali, the Acid, and Corrosive; these are the depending powers of all colours which I shall endeavour to shew in each colour in course.

FIRST, *The blue.* The blue with indigo depends on the power of the alkali, sig or urine; pearlash and potash and the lie of lime are all alkalies: so it evidently appears that indigo, although the best dye drug in the world, (except cochineal) is of no effect without the power of the alkali. There are other materials used with the indigo, but are of no other use than to support and assist the indigo: Woad will dye a blue, properly prepared, without the indigo, and indigo without the woad; so woad serves only as an assistant to the indigo. Woad is a very useful dye drug in carrying on large manufactories; but it will not answer any useful purpose in our small business. Madder is a strong drug, serving to brighten and darken the blue, which greatly assists the indigo. Wheat bran serves only to soften the water, and urine or sig prepares the dye to come to work sooner than it otherwise would. Borax is an alkali which softens all parts, and causes it to rest easy, and come to work well and soon. Blue with indigo is coloured with drugs altogether.

Prussian Blue is of a different nature; it is dependant on the power of the acid, which I shall describe hereafter. Blue with logwood is of a different nature from any other real colour. I think this is possessed of all the powers and mixed powers; with regard to logwood I have imbibed an idea that it was leading and allied to a blue, I have tried one power and another, until I have been brought to this conclusion. Madder to strengthen the logwood; allum is an acid that raises the lustre of the blue, but not sufficient of itself, it being a weak acid; verdigrease is evidently possessed of two powers, I think; it agrees with the acid and corrosive, but is most powerful as a corrosive. Sig is a weak alkali, which shows that the powers are mixed; it rouses and gives lustre to the logwood, and makes a fine blue. Thus we find the three powers may be mixed together in a real colour, although much averse to each other. Blue vitriol is possessed of two powers, acid and corrosive, and powerful in both; it has a speedy effect on logwood; and is very good in the latter part of dye, to raise, bind, and darken the colour.

In the 5th receipt I have placed the two powers as a preparation for the blue, which are in themselves in direct opposition to each other. The acid being most powerful, it will generally destroy the corrosive. Copperas is a corrosive;

allum and tartar are acids, which soften and take off the ill effects of the copperas; thus mixed together, they have a good effect; but place them in two different bodies and apply your goods, and one will destroy the other. The copperas agrees well with logwood, for almost any colour: however, for a blue, it is necessary to rince the copperas well out of your goods, otherwise the colour will be dull. Copperas being placed with the logwood, kills the nature of it, and destroys the lustre of the blue if used after the logwood. The verdigrease, sig and pearlash make the three powers compleat in this dye, only in different form and manner; which evidently shews that blue with logwood cannot be made without these powers; but when the three powers are necessarily fixed or placed in union, they must be in a feeble form; and still, if they are not properly applied, although weak and feeble, perhaps they will breed a war that will cost something before a peace can be made; so be cautious in dealing with too many powers at once, till you become well acquainted with their relative and combined strength.

PRUSSIAN BLUE.

PRUSSIAN Blue depends principally on the indigo raised by the power of the acid, and softened by the power of alkali. Oil of vitriol is a strong acid, salt and lime are alkalies; salt may be used, and answer the purpose of lime, so it evidently appears that salt is a simple alkali: these three ingredients mixed together, make a compound of blueing for Prussian blue and green. Green is no colour of itself, but is connected with two, blue and yellow, which are both dependant on the acid. Fustick is an excellent dye-wood, but is useless without the acid to raise and brighten the colour. Allum is commonly used, but tartar and aquafortis serve to raise the colour of the yellow; so green may be made very easy, the two colours being in perfect union with regard to powers. So lead them together with care till they arrive at their proper state, which is a good green.

In the 10th receipt I have admitted a little logwood and copperas, which serves to darken the green, and rather dull, &c.

In the 11th receipt, I have admitted pearlash, allum and aquafortis with the fustick. Allum and aquafortis are acids; pearlash is an alkali; the acid raises the yellow, the alkali softens and takes off the harsh parts of the acid, fits and springs the wool, to prepare it to receive the blue; the acids are binding, and the alkali the reverse.

In the 12th receipt, I have admitted wheat bran wet with vinegar. Vinegar is the principal, it being an acid, leading to corrosive, or is in greater union than any other acid with the corrosive; but when mixed with wheat bran, it is a mild acid and has quite a different effect from what it would in the natural state; and cannot be used any other way in these kind of dyes. When mixed with the bran, or otherwise, it is of a cleansing searching nature. I have admitted red tartar, which is cleansing and prepares the cloth or goods to receive a colour. Copperas serves only to darken, as I have said before.

In the 13th receipt I have admitted blue vitriol, which serves to darken and raise the lustre of the yellow.

BOTTLE GREEN.

BOTTLE Green is connected with three different colours; two as green, one as brown; the green is possessed of the quality described before, depending on the acid; the brown is assisted by the logwood, and lowered down by the power of the corrosive. The copperas would destroy the power of the acid in this dye, were it not for the verdigrease being possessed of two powers, which renders both mild and friendly.

In the 14th receipt, it is evident that blue vitriol is of two powers; as an acid it raises the yellow of the fustick, as a corrosive it darkens very rapidly with logwood, so the goods are prepared with these two powers to combine the three in one colour.

The 15th receipt is an olive green; this is a simpleness of green, and depends on the power of the acid, as green; but as brown on the corrosive; the acid going under cover of the bark, gives admittance to the corrosive; and thus the lustre of the colour is preserved from danger.

YELLOW.

YELLOW is one of the material colours, and is dependent, always, on the power of the acid, and no other; but has different subjects. Fustick is the principal subject among our dyers, and allum the principal acid. Aquafortis is very good to cleanse and prepare the colour; and it substantiates the yellow, and makes it much brighter. So the allum and aquafortis agree in all light colours; but aquafortis will not answer with a corrosive; for it is so strong an acid that it will not admit any thing of a darkening nature, as you see in taking the colour out of cloth, &c. The composition is made up of acids, and that will destroy the power of the corrosive and alkali, and all the subjects that unite with those powers; so it is evident that the acid is most powerful—for it will destroy what the others create; yet the acid may be overcome, in some of the most feeble parts, when not guarded with care by alkali and corrosive.

SCARLET RED.

SCARLET is one of the most noble colours ever made by man: cochineal is its grand and principal subject, which is the finest and best dye drug in the world. Scarlet has the most brilliant rays of all colours, which resemble the sun in the firmament and the bow in the clouds. Yet cochineal is the most simple of all dye drugs, were it not for the power of the acid and a proper connection with other subjects. The fustick and turmeric place a foundation to give lustre to the scarlet; aquafortis and argal cleanse, and raise the lustre, and make way for the cochineal to take place; yet the goods are too hard—they want softening and taking off the harsh part of the acid, which is done by wheat bran, wet with vinegar. The bran is softening, and the vinegar is an acid which is searching and cleansing.—Now the cloth is prepared to give place to the cochineal. Arsenick and armorick, are only assisting subjects; the aquafortis to keep up the life and spirits of the subjects. Thus cochineal is supported by one of the most noble and greatest powers, and is guarded by worthy subjects; and a scarlet is an ornament to kings.

The next is possessed of the same power, only the subjects are a little differently arranged.

CRIMSON.

CRIMSON has the brilliance and lustre of the rain-bow, yet is possessed of two colours; but most united with the red, with a little tincture of blue. So it is evident crimson is of no colour in itself, but is a mixture with red and blue. As red, it depends on the acid; and as blue, on the alkali. Cream of tartar, allum, and crude tartar, are all acids. Salammoniack is an alkali, and a very mild one.—Thus we find these two powers united by the help of one subject.

The other, or the next following, has a number of subjects, but dependent on the same powers.—Red, with nicaragua, is dependent on the acid, and all other reds. Dye woods are not so permanent as drugs, nor so brilliant in rays; but answer a good purpose for common use, and make very good colours. All crimsons are dependent on the powers of the acid and alkali.

MADDER RED.

MADDER is a fine drug, and may be cultivated among us, very easy. It is a tender root; and when manufactured fit for use, there are three different sorts proceeding from one root. The dyer ought to be well acquainted with the qualities of this drug. It will not admit of boiling; it kills and destroys the nature of it, (as it does all other dye-stuffs taken out of the ground.) Madder requires the softest water in the world. In order to soften the water, I use the wheat bran. But madder depends partly on two powers—when sig is used, which serves to darken and bind the red ; but brazilletto has the same effect, only the colour will be brighter—and this serves as an assistant, and the sig as an alkali, and the allum and argal as acid. Thus the madder red is dependent on the acid.

The Meroon Red has the same principal subjects, and is dependent on the same powers; and differs in nothing only it is a brighter red, and a little different in the management.

POLISHED RED.

This colour is the most independent of any colour ; not depending on any power or powers. Nutgalls is a subject with madder, but a little pearlash may be added in case it wants help ; so it appears that the alkali stands as a power, in this ; so all reds are dependent on the acids.—The crimsons and clarets are nothing of themselves, and are subject to two powers—the acid and the alkali. The subjects being differently arranged, causes the different complexions.

The power of the corrosive to destroy the power of acid.—Copperas, the strongest of all corrosives, properly prepared, will, without assistance, destroy the acid. Take cloth from acidous liquor and put it in copperas water, and it will wholly destroy the acidous power ;—and acid will destroy the corrosive, in the same manner. So it requires a mediator, when these two powers come together, to unite them, and prevent their destroying each other ; but in the mixture of colours they will require a frequent and friendly correspondence.

ORANGE COLOUR.

ORANGE colour is fine and brilliant, and

has the shades of two colours—yellow, in full; and red, in part. So orange is the union of two colours which agree in powers only.

CHERRY COLOUR.

CHERRY is a dark red, and is subject to the powers of acid and alkali; and the subjects are barwood and brazilletto—but barwood is the most depending one,though the other isnecessary.

VIOLET COLOURS.

VIOLETS are a mixture of red and blue; the red depends on the subject of brazilletto and on the power of the acid—the blue, on the subject of logwood and on the power of the alkali. Thus, in this dye, the powers and subjects agree, and by varying the powers and subjects, alter the complexions.

PINK COLOURS.

PINKS are of various colours, but this is a simple red, and is dependent on the power of the acid; Its subjects are a number, but I have laid them down as one in the receipt, and that is madder—which is the principal subject to be depended on in this colour.

FLESH COLOUR.

FLESH colour is a simple colour of red, changed from white to a small tincture of red.

This has a number of subjects, but is dependent on the power of the acid.

BROWN.

BROWN has many subjects, and of various complexions, principally dependent on the power of the corrosive; but sometimes we admit the weak power of the alkali, like the sig, &c. Brown has the greatest connection with all the colours, of any colour: for most, or the greatest part of the mixed colours, are connected with the brown, as we shall shortly shew.

CORBEAU, WITH CAMWOOD.

CORBEAU is a mixture of two colours, red and brown; these colours, in this one, dependent on two powers, and but one principal subject. The powers are an acid and corrosive; the subject, camwood and the best of dye-wood. The red depends on the oil of vitriol for an acid; to prepare and unite the bodies of the goods to take off a gray that these colours are inclining to; the blue vitriol being possessed of two powers, intercedes for the brown, supports the red and raises the lustre, which is the glory of these colours when united together; the cloth or goods, in a direct view, will be brown but when glanced by the eye or looked across, it will appear with a fine lustre of red.

The acid is a guard to the red, but that would not give admittance to the brown, were it not for the blue vitriol being of two powers, which interpose for their mutual good. Copperas, the strongest of corrosives, is harsh and fiery, and wants to be softened down notwithstanding the blue vitriol. Were it not for another assistant uniting with the corrosive, you would fail in the union of these two colours; by dissolving the copperas in vinegar, it softens the copperas; the vinegar being an acidous power, uniting with the corrosive, causes the two powers to unite. The logwood assists the camwood in completing the necessary union. Thus when these two colours, which are in opposition to each other have occasion to unite, it must be by the mediation or the subject of two powers, as I shall shew more plainly in the next place.

CORBEAU WITH NICARAGUA.

NICARAGUA, not of so spirited a nature, requires the greater assistance of the powers. This has the assistance of three powers, and has assisting subjects; the fustick, as an assisting subject, raises the lustre of the red; and yellow always depends on the acid; the blue vitriol guards the acid against the corrosive, keeps it from danger, and fits it to receive the subject of logwood; the verdigrease supports the acid, raises the lustre of the red, and unites with the corrosive; the copperas being softened by the sig, the weak power of the alkali. So by the union of the three powers, and two mixed powers, and the subjects, (the Nicaragua the chief,) the two colours are brought to an union.

CORBEAU WITH RED WOOD.

REDWOOD has spirit sufficient, but is slow

in motion, and is a feeble subject; and yet is a subject of great use: however, it requires assistance, otherwise it would fail. It is supported by the three powers, the acid and corrosive are its main dependencies; but I have placed them in different forms, as you will see by the receipts for corbeau and London brown with red-wood. The powers must support the different subjects according to the different order in which they are admitted. I have left some, deficient of the power of the corrosive, to the assistant subject logwood, and the power of the weak alkali sig; but in case the colour is not dark enough, then the dyer's judgment will call his attention to look on the receipts before mentioned, and he will see the corrosive will be admitted—the copperas or verdigrease, which is commonly best to guard the red, and powerful in darkening. Thus we find the acid and corrosive are necessary with this mixture of red and brown; and sometimes softened by the power of the alkali. The dyer will always find these colours must be supported by the power of the acid and corrosive. The acid the power of the red always; the corrosive the complete power of the brown. The reddish brown and Spanish brown are dependant on the same powers, but not altogether on the same subjects, &c.

LONDON SMOKE.

THE London Smoke is a mixture of yellow and brown. The yellow is dependant on the acid, and is the substance and life of the colour. Fustick is the principal subject for the yellow, and allum the acid, but the bark is a guard to the yellow, and is a subject in favor of all powers. The smoke is a very dark colour, bearing a little red with the yellow; thus, the butternut bark substantiates every part of these colours; the Nicaragua raises the reddish hue, the logwood assists the copperas in darkening, and the sig supports the colour in every part, and enlivens it to give place to the corrosive. Thus the three powers are united in this mixed colour, with many powerful subjects which stand well to the last.

CINNAMON BROWN.

CINNAMON colour is a mixture of three colours, red and yellow in perfect union, and is dependent on the acid; and the brown, the corrosive and alkali. So the three powers, and three subjects are united in this mixture. The smoke and liver browns are simply the same as London smoke, only differing in their subjects; the camwood and madder corresponding with the fustick, and laying a foundation for the brown. Thus the subjects will unite so perfectly well together, that they are at peace with all the powers but the the corrosive; and this binds all these subjects and unites the colours.

OLIVE BROWN.

THE Olive differs nothing more from smoke than this it is not so dark, has no hue of red, and is not depending on the alkali; but the weak alkali may be admitted, (as sig) but is dependent on the powers of acid and corrosive; and the subjects of the olives are fustick the principal; the others which are many, serve to alter the complexions and give different shades. Butternut, logwood and madder unite as to shades; the blueing gives a different shade. Thus it is left to the discretion of the dyers to make use of what form they please.

SNUFF BROWN.

SNUFF Colours are formed of three colours; dependent on the yellow for lustre, and the red and brown for the shades. The snuff colours are dependent principally on the power of the acid and corrosive, and a little on the power of the alkali; and the many different subjects have correspondence with these powers. Their union in this manner causes the difference in complexions. So as to the powers, properties and effects of these browns, they are simply all as one, but differing in complexions; I mean the smoke, the olive and snuff. It is dependent on the fustick and the acid: the red is dependent on the redwood, camwood and madder, and on the power of the alkali; the brown on the barks, the logwood, and sumac; and is dependent on the corrosive. Thus by changing the orders of the subjects and powers, the different shades may be produced in those colours; and this I have left to the discretion of the dyer.

BAT-WING, SLATE, DOVE OR LEAD, PEARL OR SILVER GREY, AND DRAB.

AS to these colours, they are a mixture of all colours, and are depending on all the powers and almost all the subjects. Some shades are very light, merely changed from white; some are dark near to a black, but as to the above colours they are stiled greys, forming various shades and complexions, from a pearl to a slate colour; the different subjects, corresponding with the powers, causes the complexions to differ. So with regard to the powers, I think I have described plainly before; the union of the subjects and colours are of so extensive a nature in these different shades, it is in vain to describe them in manner and form as I have the rest, for it would swell a volume. I have been very particular in the receipts, and given rules sufficient, and an extensive assortment of shades; but in short, they are all greys of different complexions, being of a weak and feeble make, and must be nursed with care, otherwise they will never arrive to a state of maturity.

RAVEN.

THE raven is a mixture of two colours, blue and black; black direct, and blue by the glance of the eye. Now the blue is dependent on the power of the acid and alkali, and the black on the power of the corrosive. The wheat bran softens the goods; the vinegar as an acid cleanses them and prepares them to meet the subjects, and the madder and allum rouses it up for the logwood; lying and souring gives penetration and admittance to the remaining subjects, and the corrosive power.

CROW WITH COPPERAS.

CROW colour differs not much from the raven. If any, only in form; but I think there is a difference—the crow is attended with a little brownish hue, and is dependent on the power of the corrosive, and the subject of the logwood, &c.

CROW WITH COMPOUND OF BLUEING.

THE blue part is raised with the blueing which has been described before; the black on the corrosive; the logwood the principal subject; the verdigrease intercedes for both, and unites both colours together.

CROW WITH BLUE VITRIOL.

BLUE Vitriol being connected with two powers, the acid and corrosive, forms an union with these two colours, and prepares them to meet the subject of logwood, and brings them on terms never more to part.

BLACK.

BLACK is a colour of all colours. It has but one shade, and that is the shade of darkness. Black is dependent on the power of the corrosive, and has many subjects; but logwood is the principal, the others serve as assistants to the logwood. Thus one power and one subject form the substance of this colour. There are different shades of all colours except black.

Some men, and even philosophers, have endeavoured to shew that black is not a colour;

but I shall endeavour to refute them. Black is made of materials, as any other colour; darkness is caused by materials, by the earth and the material world; by the shadow of these, darkness comes; and by the subjects of materials, white is changed to black. So men may as well argue that light is darkness, as to say that black is not a colour. Light is not darkness, nor white black; but were the light to remain with us, we should not perceive the darkness; and if we were not blessed with materials, we should not change white into colours. Light is changed by materials; the light of this world is of a nature to be changed, and white is of the same substance, depending for its changes on materials of dye-stuff; by our faculties we use them, and obtain the desired effect which God in his wisdom has designed. Blue, yellow, red, brown and black are made of materials; they are all colours, and are all of equal rank, formed from white; yet black is most powerful, for that may be made to overshadowall other colours, and cause darkness to reign over them all. So it is evident that black is a colour of all colours. But black and white mixed together is no colour. If light and darkness were mixed together, we should have neither: the God that made the world separated the light from the darkness; so in like manner he has given us materials, and a faculty to use them, to change white to black. Thus we find that black is a colour.

It is said that orange and violets are colours, but they are not in themselves so, but are compounded of colours. No mixture can be a real colour.

Having endeavoured to give you my ideas of the properties and effects of colours; I request to be read with candor, and hope to be of some benefit. If I have committed errors, I wish they may be corrected for the public abvantage.

COLOURING SILK.

SILK is of a nature different from wool, cotton, and linen : it is of a deadly nature : however, the most of preparations for dying woollen will answer for silk, only the dye requires to be stronger. It has also such an union with cotton and linen, that most of these preparations will answer for either. So it appears that silk is of a substance between wool, cotton and linen, and it unites with them as to colours, &c.

DYING COTTON AND LINEN.

COTTON and Linen are of a cold and deadly nature, and require different preparations and management in colouring. It is the best way in colouring cotton and linen, to have the dye cold ; they being of so cold a nature. As to the colours of cotton and linen, I shall say but little : As to the powers, the principal is the corrosive, the next the alkali, and sometimes the acid ; which you will see by the receipts. The subjects are many, but the grand subject is nutgalls ; the others are so numerous, I shall not mention them now. I have endeavoured to explain them explicitly in the receipts for cotton and linen, and think it needless to mention them again here. As to the powers and the union of the subjects, they have been explained before and the best way is to examine the rules for improvements, and follow the receipts close in proper order, and I presume to say they will have the desired effect, in all colours and shades.

OBSERVATIONS ON THE PRESENT SITUATION OF THE DYER'S BUSINESS.

Observation First.

WE think ourselves masters of our business before we are, and undertake to do that we know nothing of. By this our business is ruined, our customers imposed upon, and our country impoverished; this is the present situation of our business.

Observation Second.—Those impostors injure their fellow-functioners as well as the public, by discouraging manufactories. Finding they fail of their intentions, they begin to encourage their customers by promising to do better, and to work very cheap; by these impostors, people are deluded, and their goods not unfrequently ruined. With the customer, who knows nothing of dressing, cheapness is every thing. The workman who is a complete master of his business is often compelled to regulate his prices of work by the charges of those who are ignorant of the trade; consequently the work is slighted, or the mechanic cannot obtain a living; and the employer is a loser in the end, as the goods are badly finished, or perhaps entirely ruined.

Let those who practise in a business make themselves masters of it; then fair and just prices may be obtained for their labor, and the employer will be better satisfied, and real justice be done him.—Thus our manufactories would be increased: The interests of the employer and employed would both be enhanced; they are inseparable: selfishness counteracts its own views; the injustice we do our country, we do to ourselves.

As a nation we can never be really independent, until we become our own manufacturers of articles of the first necessity. To arrive at this desirable point ought to be our constant endeavour; and every real patriot will use his exertions, not only in word but in deed, to hasten the period.

OBSERVATIONS ON MANUFACTURING CLOTH.

TO make fine Broadcloth,take your wool and sort it carefully; take the shortest and finest of the wool, leaving no coarse locks with it; then break the wool all together, and card it into rolls by one person or machine, then spin well the filling cross handed; give it a good twist, but not hard so as to be wirey; let it all be twisted alike, and spun by one person, then let it be well wove, with the threads closed together, but not too hard. Then take the long wool, and have it combed into worsted; have it well spun, twisted well, and wove firm. Thin cloths depend on the twisting and firm weaving; but the other, after it is well manufactured, depends on the fulling to close and make it firm, and on the dressing for beauty. The cloth if well manufactured, well fulled, and well dyed and dressed, will appear equal to any imported cloths; but if not well manufactured, it will not be handsome.

If you have coarse wool and fine mixed together, it cannot be a fine piece; if it is not broke and carded together, it willnot work well; it is liable to be streaked,and pucker or cockle in the mill. If not well spun, or if spun by two hands, it will have the same effect; and if two weave on one piece, one thick and the other thin, it will cause it to puckeror cockle.

With proper care and attention in the manufacturing and dressing of cloths, we may equal any in workmanship and beauty, and afford them one third cheaper than those imported.

APPENDIX

TO THE

DYER'S COMPANION.

INTRODUCTION.

THE author of the Dyer's Companion thinks it his duty, in gratitude to the professors in that noble art, to subscribe his hearty thanks for their approbation of, and encouragement given to that work, in this and every part of America. It is well known by that body of people, and felt by some, that the price for dying woollen goods has been much reduced of late. Many circumstances having determined me, long since, to acquire all possible knowledge in the practice of dying; in my first edition I had not the advantage of any author to assist me, in compiling or improving this noble art, it is a work of my own study and improvements; I have of late had the advantage of some authors, showing the general practice of dying all over Europe, in England, France, Germany, &c. which I shall in this edition give, a general plan of dying, together with my own improvements from my small practice, and some observations on the properties and effects of dyes and dye stuff, the modern and ancient forms, as to the use of them. I am therefore constrained, once more to recommend a strict inquiry into the original quality of all the drugs they use, that thereby, if possible, they may discover some of the hidden advantages that may justly be expected therefrom. I am astonished that no artist has ever attempted to improve this most ingenious art of chymical principles, I began the work in hopes that my master-piece would induce some Artist to undertake its improvement, but in vain do I expect it.

Of flowers.—Among the infinite variety of colours, which glow in the flowers of plants, there are very few which have any durability, or whose fugitive beauty can be arrested by art, so as to be applied to any valuable purpose; the only permanent ones are the yellow, the red, the blue, and all the intermediate shades of purple; crimson, violets, &c. are extremely perishable. Many of these flowers lose their colour on being dried, especially if they are dried slowly; the colours of them all perish even in the closes vessels, the quicker they are dried, and excluded from the air, the longer they retain their beauty. The colouring matter, extracted and applied on certain bodies, is still more perishable; oftentimes it is destroyed in the hand of the operator. The colour of blue flowers is extracted by the infusion of water; but there are some from which this gains only a redish or purplish hue. The red flowers readily communicate their own red colour to water, there is no exception among those that have been tried; rectified spirits give a brighter, but paler than the water infusion. The colour of yellow flowers, are in general durable; many of them are as much so perhaps as any of the native colours of vegetables. The colour is extracted both by water & spirits; the watry infusions are the deepest. Neither acid nor alkalis alter the species of colour, though both of them vary the shades; acid, making it paler, and alkalis deeper: allum likewise considerably heightens it, though not so much as alkalis. Woollen silk impregnated with a solution of allum and tartar, receives a durable yellow. Some of these flowers were made use of by the German dyers.

Of fruits.—The red juices of fruits are, red currants, mulberries, elder berries, morello, black cherries, &c. gently inspissated to dryness, dissolve again; almost totally in water, and appear nearly of the same red colour as at first. Rectified spirits extracts the tinging particles, leaving a considerable portion of muilaginous water undissolved, yet the spirituous tincture proves of a brighter colour than the watry. The red solutions and the juices themselves, are sometimes made dull, and sometimes more florid, by acid, and generally turned purplish by alkalis. There are a great number of fruits of different shades which I shall not attempt to enumerate or describe. But to prove, the proofs of colouring vegetables is by varying them with spirits, acids and alkalis, saline, volatile and corrosive liquors. When I make use of the word corrosive it is intended to imply, to absorb, to destroy, to reign king like black, or to change any shade, and destroying the lustre, or lustre of the colour; it is one of the reigning powers, let the substance or quality be what it will.

Of Plants.—The blue and green colours of leaves and plants, have more or less fusibility; we cannot claim in the mineral and animal kingdoms, no substitute for blue, except (Prussian blue,) one which has been introduced by some dyer's as a colouring substance, and the vegetables yield but two, which are both produced from the leaves of plants: indigo and woad. For yellow, there are seven different plants besides woad and barks, which will be spoken of hereafter. The leaves of many kinds of herbs and trees give a yellow dye to wool or woollen cloth, that has been boiled in a solution of allum and tartar; wild in particular affords a fine yellow, and is commonly made use of for this purpose by the dyer's, and cultivated in large quantities in some parts of England. There is no colour for which we have such a variety of materials as for yellow; Mr. Hellot observes, that all leaves, barks and roots, which on being chewed discover a slight astringency, as the leaves of the almond, peach, and pear trees, ash bark, the roots of wild patience, &c. all yield durable yellows; the brightness will be according to your preparation of allum and tartar, and length of time you boil your dye, &c. If we were sensible of the double advantage that might be acquired in the use of many of our vegetable drugs, which must be first grounded on chymical experiments in miniature, which will be a certain rule to the practice at large, I am certain you would not rest till you had made some improvement. For experiment, after you have been dying with that resinous drug sanders, when emptying the vessel, you take up a handful, dry it and digest it in a phial with some pure spirits of wine, and it will afford you an excellent red; water being insufficient to dissolve the resin, and set out the prime part of the colour. Many others may be discovered if an unwearied attention was paid.

Of dying in general. — If I ask a dyer what ingredients compose a black? the answer will be this, logwood, sumac, galls, bark, and copperas; and if he knows it, he will add in his last dip a little ashes or argal. If I ask him, which of these drugs contain an acid, which an aleotic, and which a neutral quality, he cannot give me an answer; so you see he knows the effect, but is a stranger to the cause, and every thing else separate from fact and custom. What a pity it is that men will not search things to the bottom, when they might be able to find out the causes of miscarriages, for which goods are frequently thrown aside to be dyed other colours, greatly to the dyer's loss. In conversing with a sensible dyer I simply asked him what part does logwood act in the black dye; the honest man answered, it helps to make it black; no other proof was wanted but to follow the old round; but the reader by now thinks it time to be informed of the business of logwood, which is, (if used in a right proportion,) to soften the goods, and give a body and lustre to the colour. Logwood being possessed of a most excellent astringent quality, fixes itself in the pores of the goods, and gives them a velvet-like feel and gloss.

Some will object to this assertion, and say, but our blacks have not that velvet-like feel and gloss; true, sir; but don't you know the reason; you dye your black without scowering your goods, forgetting, or not knowing, that when the goods enter the boiling dye liquor, they grow harsh and the oil contained in them forms a sort of resin, which becomes as fixed as if it was pitch or tar, this is one great reason why black is so liable to soil linen because the dye in some sense is held in an outside or superficial state; it is not possible these goods should finish soft like velvet, or shine like a ravens feather! No, on the contrary they spoil the press papers and come out stiff and hard like buckram, (not velvet,) no greater cause can be assigned for it than that of not scowering; this is the reason of the great difference, so much spoken of, between the London blacks and those dyed in America; if the American dyers would take the same pains as the Londoners do, I think they would excel in fact, if not in name, and therefore let the American dyers be equally tight and clean in their performances, and there is nothing to prevent their superiority. Many will censure and despise this, for no other reason than because they cannot see into it, nor will they be at any pains to learn and improve their talents; they seem rather to choose the old round, having no spirit or courage to improve, but content with each knowing the other's method, without striving to excel, and discover a more complete and less expensive way of working and using the drugs to the best advantage. I know not how men can sit still when there is more to learn; let it not be said of you as one of old, he lived and died and did nothing; perhaps he worked with his hands but his head was asleep, and when dead his memory was no more; sure it is, the invitation I have to write and publish this small treatise on dying is not so much to please others, or to show any thing I have is capable of the name of parts, but to communicate my good wishes for improvements to my brethren the dyers, and to show my willingness to help to perfect one of the most useful arts in the world.

There are very few arts so expensive as that of dying; and although those principal commodities, clothing and furniture, receive their chief improvement and value therefrom, it is nevertheless very far from being brought to perfection. A long practice, sound judgment, and great attention, will form a good and expert dyer. Many dyers can work with success in a number of colours only which depend on each other, and are entirely ignorant of the rest, or have but a very imperfect idea of them.

A philosopher, who studies the art of dying, is in some measure astonished at the multiplicity of new objects which it affords; every step presents new difficulties and obscurities, without hopes of any instruction from the common workmen, who seldom know more than facts and custom. Their manner of explaining themselves, and their common terms, only afford more darkness, which the uncommon and often useless circumstances of their proceedings render more obscure.

Before we enter into the particulars of dying wool, it is necessary to give an idea of the primary colours, or rather of those which bear this name by the artist; for it will appear by reading the celebrated works of Sir Isaac Newton on Light and Colours, that they bear no affinity with those which the Philosophers call by that name. They are thus named by the workmen, because by the nature of the ingredients of which they are composed, they are the basis from whence all others are derived. This division of colours, and the idea which I intend to give of them, are also common to the different kinds of dying.

The five primary colours are blue, red, yellow, brown and black. Each of these can furnish a great number of shades, from the lightest to the darkest; and from the combination of two or more of these different shades, arise all the colours in nature. Colours are often darkened, or made light, or considerably changed, by ingredients that have no colour in themselves; such are the acid, the alkalis, and the neutral salts, lime, urine, arsenic, allum, and some others; and in the greatest part of dyes, the wool and woollen goods are prepared with some of these ingredients which of themselves give little or no colour. It may easily be conceived what an infinite variety must arise from the mixture of these different matters, or even from the manner of using them; and what attention must be given to the minutest circumstances, so as perfectly to succeed in an art so complicated, and in which there are many difficulties.

It is not needful to be very particular in describing the utensils of a dye-house, as they are commonly known; this work being designed for the experienced dyer. A dye-house should, however, be erected on a spacious plan, roofed over, but admitting a good light, and as nigh as possible to a running water, which is very necessary, either to prepare the wool before it is dyed, or to wash it afterwards. The coppers should be set at the distance of eight or ten feet, and two or more vats for the blue, according to the quantity of work that is to be carried on.

The most important point in dying the primitive blue is to set the vat properly at work, and conduct her till she is in a state to yield her blue. The size of the woad vat is not fixed, as it depends upon necessity or pleasure. A vat containing a hogshead, or half that quantity, has often been used with success, but then they must be prevented by some means from cooling too suddenly, otherwise these small vats will fail.

Another kind of vat it prepared for blue: this is called the indigo vat, because it is the indigo alone that gives it the colour. Those that use the woad vat do not commonly use the indigo one.

There are two methods of dying wool of any colour; the one is called dying in the great, the other in the lesser die. The first is done by means of drugs or ingredients that procure a lasting dye, resist the action of the air and sun, and are not easily stained by sharp or corrosive liquors. The contrary happens to colours of the lesser dye. The air fades them in a short time, more particularly if exposed to the sun; most liquors stain them, so as to make them lose their first colour. It is extraordinary that, as there is a method of making all kinds of colours by the great dye, the use of the lesser should be tolerated; but three reasons make it difficult, if not impossible, to prevent this practice.

1st, The work is much easier. Most colours and shades which give the greatest trouble in the great, are easily carried on in the lesser dye.

2d, Most colours in the lesser are more bright and lively than those of the great.

3d, For this reason, which carries more weight, the lesser dye is carried on much cheaper than the great. This is sufficient to determine some men to do all in their power to carry it on in preference to the other. Hence it is that the true knowledge of chymistry, to which the art of dying owes its origin, is of so much use.

It may be observed, that all lasting colours are called colours of the great, and the others of the lesser dye. Sometimes the first are called fine, and the latter false colours; but these expressions are equivocal, for the fine are sometimes confounded with the high colours, which are those in whose composition cochineal enters; therefore, to avoid all obscurity I shall mention them distinctly and separately in their places hereafter.

Experiments, (which are the best guides in natural philosophy as well as arts) plainly shew that the difference of colours, according to the foregoing distinction, partly depends on the preparation of the subject that is to be dyed, and partly on the choice of the ingredients which are afterwards used to give it the colour. I therefore think it may be laid down as a general principle, that all the invisible process of dying consists in dilating the pores of the body that is to be dyed, and depositing therein particles of a foreign matter, which are to be detained by a kind of cement which prevents the sun or rain from changing them. To make choice of the colouring particles of such a durability that they may be retained, and sufficiently set in the pores of the subject opened by the heat of boiling water, then contracted by the cold, and afterwards plaistered over with a kind of cement left behind with the salt used for their preparation, that the pores of the wool or woollen stuff ought to be cleansed, enlarged, cemented and then contracted, that the colouring atom may be contained in a lasting manner.

Experiments also shew that there is no colouring ingredient belonging to the great dye which has not more or less an astringent and precipitant quality. That this is sufficient to separate the earth, of the allum; this earth joined to the colouring atoms, forms a kind of lacque, similar to that used by the painters, but infinitely finer. That in bright colours, such as scarlet, where allum cannot be used, another body must be substituted to supply the colouring atoms (block-tin gives this basis to the scarlet dye.) When all these small atoms of earthy coloured lacque have insinuated themselves into the pores of the subject that is dilated, the cement which the tartar leaves behind serves to masticate these atoms; and lastly, the contracting of the pores, caused by the cold, serves to retain them.

It is certain that the colours of the false dye have that defect only because the subject is not sufficiently prepared; so that the colouring particles being only deposited on its plain surface, it is impossible but the least action of the air or sun must deprive them of part if not of the whole. If a method was discovered to give to the colouring parts of dying woods, the necessary astriction which they require, and if the wool at the same time was prepared to receive them, (as it is the red of madder) I am convinced, by thirty experiments, that these woods might be made as useful in the great, as they have hitherto been in the lesser dye.

What I have said shall be applied in the sequel of this treatise, where I shall shew what engaged me to use them as general principles.

I should have been glad to have seen a work of this sort, (knowing the great need there is of a chymical understanding of this art) signed by the name of some person of distinction, to have given it a better face. I dare nor flatter myself to have brought it to its last perfection, as arts daily improve, and this in particular; but I hope some acknowledgment will be due to me for bringing this matter a little further out of that obscurity in which it has laid, and for assisting the dyers in making discoveries to help to perfect this most useful art.

I shall now proceed to examine the five primary colours above mentioned, and give the different methods of preparing them after the most solid and permanent manner.

The materials of which cloths are made, for the most part are naturally of dull and gloomy colours. Garments would consequently have had a disagreeable uniformity, if this art had not been found out to remedy it, and vary their shades. The accidental bruising of fruits or herbs, the effect of rain upon certain earths and minerals might suggest the first hint of the art of dying, and of the materials proper for it. Every climate furnishes man with ferruginous earths, with boles of all colours, with saline and vegetable materials for this art. The difficulty must have been to find the art of applying them. But how many trials and essays must have been made, before they found out the most proper methods of applying them to stuffs, so as to stain them with beautiful and lasting colours? In this consists the principal excellence of the dyer's art, one of the most ingenious and difficult which we know.

Dying is performed by means of limes, salts, waters, lies, fermentations, macerations, &c. It is certain that dying is very ancient. The Chinese pretend that they owe the discovery of it to Hoan-ti, one of their first sovereigns.

One of the most agreeable effects of the art of dying, is the diversifying the colours of stuffs. There are two ways by which this agreeable variety is produced, either by needle-work with threads of different colours, on an uniform ground, or by making use of yarn of different colours in the weaving.

The first of these inventions is attributed to the Phrygians, a very ancient nation; the last to the Babylonians. Many things incline us to think that these arts were known even in the times of which we are now treating. The great progress these arts had made in the days of Moses, supposes that they had been discovered long before. It appears to me certain, then, that the arts of embroidery or weaving stuffs of various colours were invented in the ages we are now upon. But I shall not insist on the manner in which they were then practised, as I can say nothing satisfactory upon that subject.

Another art nearly related to that of dying, is that of cleaning and whitening garments, when they have been stained and sullied. Water alone is not sufficient for this. We must communicate to it by means of powders, ashes, &c. that detersive quality which is necessary to extract the stains which they have contracted. The ancients knew nothing of soap, but supplied the want of it by various means. Job speaks of washing his garments in a pit with the herb borith. This passage shows that the method of cleaning garments in these ages, was by throwing them into a pit full of water, impregnated with some kind of ashes; a method which seems to have been very universal in these first times. Homer describes Nausicaa and her companions washing their garments, by treading them with their feet in a pit.

With respect to the herb which Job calls borith, I imagine it is salworth. This plant is very common in Syria, Judea, Egpyt, and Arabia. They burn it, and pour water upon the ashes. This water becomes impregnated with a very strong lixivial salt proper for taking stains or impurities out of wool or cloth.

The Greeks and Romans used several kinds of earths and plants instead of soap. The savages of America make a kind of soap-water of certain fruits, with which they wash their cotton-beds and other stuffs. In Iceland the women make a lie of ashes and urine. The Persians employ boles and marls. In many countries they find earths, which, dissolved in water, have the property of cleaning and whitening cloth and linen. All these methods might perhaps be practiced in the primitive ages. The necessities of all mankind are much the same, and all climates present them with nearly the same resources. It is the art of applying them, which distinguishes polite and civilized nations from savages and barbarians.

I shall leave all to itself, and to every man liberty to approve or disapprove as he pleases, and however they determine the author will not be much troubled, for he is certain no man can have a lighter esteem for him, than he has for himself; he however, will be well pleased if any man shall find benefit by what he has written. If any should alledge a general opposition, that to the author, will be no privating argument; he does not plead the importunity of friends, for the publication of this; if it is worthy it needs no apology, if not, let it be despised; and I remain the same friend to trade.

ELIJAH BEMISS.

APPENDIX, &c.

CHAPTER I.

ON BLUE DYING.

BLUE among all colours is the most difficult to set up and manage, it is one of the five material or primitive colours. In the preceding work it is to be observed, that the rule given is calculated for cloth generally: in this I shall give the process for wool, for that is to be preferred, and the most sure and the only way blue ought to be coloured, except very coarse cloth; and further, I have given in the preceding, receipts for blue, for the good and false dye, or the greater and the lesser dye; in the good dye we are not furnished but two drugs that give a permanent blue, and they are, indigo and woad, or pastel, and these light and fleaty substances, were it not for the power of fixed alkalis, which rouses and gives life to the colouring atom or substance, and separates it from the earth it contains; I shall leave this for further explanation.

The lesser dye is obtained from logwood with assisting subjects, and depends on the powers of the alkali and acid, as you may see in my preceding observations on blue with logwood; it is further to be observed, in the preceding observations on the properties and effects of dyes and dye stuff, that I have classed the colours into five material or primitive colours; which are, blue, yellow, red, brown, and black; all other shades are depending on these as their mothers or princesses, and these five colours are depending on three monarchial powers, which have but little or no alliance with each other, except it is by the intercourse of some neutral power. The names of the powers are the alkali, acid, and corrosive, and all their subjects rest mild and easy under them, and have a friendly correspondence and never are at variance, except it is by the interference of the powers; when it happens, there can be no peace or negociation only by the assistance of a neutral power. I shall give further explanations on this subject in the sequel; showing the connexion of colours by twos, and by threes, and their dependencies with the dying subjects and the subjects to be dyed.

To return to the blue; 1st. it is necessary to pay some attention to the vat, and utensils used in blue dying. 2nd. The explanation on the articles used in the blue dye, how prepared, and its effects, &c. 3d. Give a brief account by way of receipts, of the modern forms as practised in general in the largest manufactories of America, and the general practice in England and France, by the most noted dyers in Europe, &c.

The vat and utensils used in blue dying.—The vat must be in size and proportion as your business requires, from eight to twenty-four barrels; the fashion and forms are various according to fancy, but I shall point out the modes most applicable, and easiest to manage and despatch business. The best and cheapest way to make a vat; have the bottom of cast iron about two and a half feet deep, with a flange on the outside about four inches from the top, then raise it to five or five and a half feet, with staves made of pine plank two inches thick, hooped with iron hoops and fastened to the upper edge of the iron kettle; when thus prepared, place it in the dye-house, where it will be the most advantageous to work at, set it with convenience for heating, with a flew raised with brick to keep the dye at a proper heat, &c. Some have them of lead and have them set, but it is expensive and liable to melt and burst; others use copper caldrons, which ought not to be admitted, for the alkali corrodes the copper and has a bad effect on the dye. The next thing necessary is a large iron boiler, that will contain half as much as your vat, set for the convenience of heating near the vat, for the purpose of setting and recruiting your dye, and immersing your goods in warm water; and a large tub that will hold the remainder of your dye. The next thing necessary is a small iron kettle, say about a barrel, set convenient for heating, for a preparation kettle to dissolve potash, &c.; another kettle is necessary, say the size of six gallons, for the purpose of grinding indigo with two eighteen pound cannon balls; the form, have the bottom rounding that the indigo may settle under the balls, and the point of the standard placed in the centre with a cross to turn the balls. The next, a copper ladle with a long handle, to dip to the bottom of the vat, to hold two gallons; a copper skimmer with a handle sufficient to reach to the bottom, to take up the ground, say eighteen inches diameter; a smaller one to take off the flury or head, and a small tub to contain it. The next thing necessary for dying wool, is a net sufficient to contain the wool, and strung with a cord the width of the vat, and its depth within two feet of the bottom, but not so as to touch the ground. There is another form of vat and utensils used for blue dying, explained in the preceding work in receipt No. 1, for dying cloth; the rake, the jack for wringing, the screen, the handlers, the folding and cooling board, &c. &c. The cold indigo vat with urine, does not require to be set in the ground, neither a flew around it, but set in the dye-house as is most convenient to work at. All woollen blue dyes require to have a tight cover, and clouted with cloths to prevent the evaporation of the volatile fluid. The cotton vats are set quite different if worked cold, which I shall describe hereafter.

The method of preparing goods for blue; and an explanation of the dye stuffs, how prepared and its effects.

WHEN the vat is once prepared and come to work, the dying of wool or stuffs is easy. Wet them well in clear warm water, with one quarter of a pound pearlash to every 40 pound of wool, wringing and dipping them in the vat, and keeping them in more or less time, according as the colour is required in shade. From time to time the stuff is aired; that is, taken out of the vat and wrung, so that the liquor may fall back into the vat, and exposed a little to the air, which takes off the green in one or two minutes; for let what vat soever be used, the stuff is always green at its coming out, and only takes the blue colour in proportion as the air acts upon it. It is also very necessary to let the green go off before it is returned into the liquor to receive a second shade, as being then better able to judge of its colour, and know if it is requisite to give what is called one or several turnings.

It is an ancient custom among dyers to reckon thirteen shades of blue from the deepest to the lightest. Although their denominations be somewhat arbitrary, and that it is impossible exactly to fix the just passage from one to the other, I shall notwithstanding give the names. They are as follow, beginning with the lightest: milk-blue, pearl-blue, pale-blue, flat-blue, middling-blue, sky-blue, queen's-blue, turkish-blue, watchet-blue, garter-blue, mazareen-blue, deep-blue, and very deep or navy-blue.

These distinctions are not equally received by all dyers, nor in all provinces, but the most part are known; and it is the only method that can be taken to give an idea of the same colour, whose only difference is in being more or less deep.

It is easy to make deep blues. I have already said, that to effect this, the wool or stuffs are to be returned several times into the vat; but it is not so in respect to light blues; for when the vat is rightly come to work, the wool can seldom be left in short time enough, but that it takes more than the shade required. It often happens when a certain quantity of wool is to be dipped, and that it cannot all be put in at the same time, that what goes in at first is deeper than the other. There are some dyers who, to obviate this inconveniency in making very light blues, which they call milk and water, take some of the liquor of the indigo vat, and dilute it in a very great quantity of lukewarm water; but this method is a bad one, for the wool died in this mixture has not near so lasting a colour as that dyed in the vat; as the altering ingredients which are put into the vat with the indigo, serves as much to dispose the pores of the subject which is dipped in, as to the opening of the colouring fecula which is to dye it, their concourse being necessary for the adhesion of the colour. The best method of making these very light blues, is to pass them either in a woad or indigo vat, out of which the colour has been worked, and begins to cool. The woad vat is still preferable to that of the indigo, as it does not dye so soon.

The blues made in vats that have been worked are duller than the others; but they may be pretty sensibly roused by passing the wool or stuffs in boiling water. This practice is even necessary to the perfection of all blue shades; by this the colour is not only made brighter, but also rendered more secure, by taking off all that is not well incorporated with the wool; it also prevents its spoting the hands or linen, which commonly happens, and the dyers, to gain time, neglect this precaution. After the wool is taken out of the warm water, it is necessary to wash it again in the river, or at least in a sufficient quantity of water for the carrying off all the superfluous loose dye.

The best method to render the blue dye brighter, is by filling them with a thin liquor of melted soap, and afterwards cleansing them from the soap by warm water, and, if convenient, by rinsing them in an old cochineal liquor. This method is to be taken with deep blues; but if the same was taken with very light blues, they would lose their bright blue lustre, and incline to grey.

I hope to have removed all difficulties on the preparation of blue, and in the method of dying it. Some dyers, for the sake of gain, spare the woad and indigo, and use for blue, archil-log-wood, and brazil: this ought to be expressly forbid, though this adulterated blue is often brighter than a lasting and legitimate blue. This is to be noticed in the receipts, treating on the lesser die.

I shall now explain the theory of the invisible change of the blue dye. This colour, which I shall here only consider in relation to its use in the dying of stuffs of what kind soever, has hitherto been extracted only from the vegetable world, and it does not appear that we can hope to use in this art the blues the painters employ: such are the Prussian blue, which holds of the animal and mineral kind* ; the azure, which is a vitrified mineral substance ; the ultramarine, which is prepared from a hard stone ; the earths that have a blue colour, &c. These matters cannot, without losing their colour in whole or in part, be reduced into atoms sufficiently minute, so as to be suspended in the saline liquid, which must penetrate the fibres of the animal and vegetable substances of which stuffs are manufactured ; for under this name linen and cotton cloths must be comprehended, as well as those wove of silk and wool.

Hitherto we know but of two plants that yield blue after their preparation : the one is the isatis or glaustum, which is called pastel in Languedoc, and woad in Normandy. Their preparation consists in a fermentation continued even to the putrefaction of all the parts of the plant, the root excepted ; and consequently in the unfolding of all their principles into a new combination, and fresh order of these same principles, from whence follows an union of infinite fine particles, which, applied to any subject whatever, reflects the light on them very different from what it would be, if these same particles were still joined to those which the fermentation has separated.

The other plant is the anil, which is cultivated in the East and West Indies, out of which they prepare that fecula that is sent to Europe under the name of indigo. In the preparation of this plant the Indians and Americans, have found out the art of separating only the colouring parts of the plant from the useless ones ; and the French and Spanish colonies have imitated them,

* 1748, Mons. Macquer, of the Royal Academy of Sciences, found the means of using the Prussian blue to dye silk and cloth, in a blue whose brightness surpassed all the blues hitherto known.

and thereby made a considerable increase of commerce.

That the indigo, such as is exported from America, should deposite on wool or stuffs the colouring parts required by the dyer, it is infused several ways, the processes of which will be given in the sequel. They may be reduced to three; the cold indigo vat may serve for thread and cotton; those that are made use of hot, are fit for stuffs of any kind whatever.

In the cold vat, the indigo is mixed with pearlash, copperas or green vitriol, lime, madder, and bran. The hot vats are either prepar-with water or urine; if with water, pearlash or potash, and a little madder must be added; if with urine, allum and tartar must be joined to the indigo. Both of these vats, principally intended for wool, require a moderate degree of heat, but at the same time strong enough for the wool to take a lasting dye, I mean such as will withstand the destroying action of the air and sun, the proof of dyes.

I have prepared, as I said before, these three vats in small, in cylindrical glass vessels, exposed to the light, in order to see what passed before the infusion came to a colour, that is whether it was green beneath the flurry at the surface, which is a sign of internal fermentation. I have said that the green colour of the liquor is a condition absolutely essential, and without which the colour the stuff would take would not be a good dye, and would almost entirely disappear on the least proofs.

I shall now give a description of the cold indigo vat in small, for the changes are much better seen in her, and for this reason, that what happens in the two others is not very essentially different. It is proper to take notice, that what I shall call *part*, in this observation of experiments, is a measure of the weight of four drachms, of all matter either liquid or solid, and that it will be this quantity that must be supposed, each time that I use that word in the detail of these experiments.

I put three hundred parts of water into a vessel, containing five hundred and twelve, or eight quarts, in which I dissolved six parts of copperas, which gave the liquor a yellow dye. Six parts of potash were also dissolved by themselves in thirty-six parts of water. The solution made, I digested in it six parts, or three ounces, of indigo of St. Domingo well ground; it was left over a very gentle fire three hours. The indigo swelled, and taking up a larger space, rose from the bottom of this alkaline liquor, with which it formed a kind of thick syrup, which was blue. This was a proof that the indigo was only divided, but not dissolved; for had its solution been perfect, that thick liquor would have been green instead of blue; for all liquor that has been tinged blue by a vegetable of any kind, grows green on the admixion of an alkaline salt, either concrete or in a liquid form, whether it be a fixed or volatile.

From hence the reason is discovered why indigo does not dye a stuff of a lasting blue when its liquor is not green; for its solution not being complete, the alkali cannot act upon these first elementary particles; as for example, it acts on the tincture of violets, which is a perfect solution of the colouring parts of those flowers, which it turns green in an instant, and on the first contact.

I poured this thick blue liquor into the solution of vitriol, and after well shaking the mixture, I added six parts of lime that had been slacked in the air; it was cold weather when this experiment was made; the thermometer was at two degrees under the freezing point, which was the cause that this was near four days coming to a colour, and the fermentation, which must naturally ensue in all vitriolic liquor, where an alkaline salt has been put in, such as potash, and an alkaline earth, was carried on with so much slowness that very little scum appeared on the surface of the liquor. In a hot season, and by making use of lime newly calcined, these kind of vats are sometimes fit to dye in four hours.

Each time I stirred the mixture with a spatula, I observed that the iron of the vitriol or copperas was the first that precipitated to the bottom of the vessel, and that the alkaline salt had precipitated it to join itself to the acid. Thus in this process of the cold indigo vat, a tartar of vitriol after the manner of Tachenius is formed; whereas by the common method of preparing this neutral salt, the acid of vitriol is poured on a true alkaline salt, such as potash. This again is a circumstance that leads insensibly to the theory of the good dye. I desire the reader to take notice of this, as it will occur in the sequel of this observation, as well as in other chapters.

The earthy parts of the lime precipitate next after the iron; they are easily distinguished by the whiteness, which are yet difficult to distinguish when the colouring parts of the indigo are sufficiently loosened. In short, under this white earth the fecula of the indigo deposites itself, and by degrees rarifies in such a manner, hat this substance, which the first day was only the eighth of an inch above the precipitated lime, rose insensibly within half an inch of the surface of the liquor, and the third day grew so opaque and muddy, that nothing further could be distinguished.

This rarefaction of the indigo, slow in winter, quick in summer, and which may be accelerated in winter by heating the liquor to fifteen or sixteen degrees, is a proof that a real fermentation happens in the mixture, which opens the little lumps of indigo, and divides them into particles of an extreme fineness; then their surfaces being multiplied almost *ad infinitum*, they are so much the more equally distributed in the liquor, which deposits them equally on the subject dipped in to take the dye.

If fermentation comes on hastily, or in a few hours, whether on account of the heat of the air, or by the help of a small fire, a great quantity of flurry appears; it is blue, and its reflection they have also named coppery, because the colours of the rainbow appear in it, and the red and yellow here predominate; however this phænomenon is not peculiar to indigo, since the same reflection is perceived in all mixtures that are in actual fermentation, and particularly in those which contain fat particles blended with salts, urine, soot, and several other bodies put into fermentation, show on their surface the same variegated colours.

The flurry of the indigo vat appears blue because exposed to the external air, but if a small portion of the liquor which is under it be taken up with a spoon, it appears more or less green in proportion as it is filled with colouring particles. In the course of this observation, I shall show the reason of this difference, or, at least, a probable explication of this change of blue, which, as I have said before, is absolutely necessary for succeeding in the process described.

When the vat is in this state, it has already been said that cotton, thread, cloths wove from them, &c. may be dyed in her, and the colours which they take are of the good dye; that is this cotton and thread will maintain them, even after remaining a suitable time in a solution of white soap, actually boiling. This is the proof given them preferable to any other, because the linen and cotton cloths must be washed with soap when dirty.

Though the indigo liquor which is in this state can make a lasting dye without the addition of any other ingredients; the dyers who use this cold vat add, as in the other hot vats, a decoction of madder and bran in common water run through a sieve; this is what they call *bever*. They put madder to insure, as they say, the colour of the indigo, because this root affords a colour so adhesive that it stands all proofs; they put the bran to soften the water, which they imagine generally to contain some portion of an acid salt, which, according to their opinion, must be deadened.

This was the opinion of the French dyers against indigo in the days of Monsieur Colbert; and as this minister could not spare time to see the experiments performed in his presence, on the foundation of this report, he forbade indigo to be used alone. But since the government has been convinced, by new experiments made by the late Mr. Dufay, that the stability of the blue dye of this ingredient was such as could be desired; the new regulation of 1737, licences the dyers to use it alone, or mixed with woad; so that if they continue to use the madder, it is rather because this root giving a pretty deep red, and this red mixing with the blue of the indigo, gives it a tint which approaches the violet, and also a fine hue.

As to the bran, its use is not to deaden the pretended acid salts, but to disperse throughout a quantity of sizey matter; for the small portion of flour which remains in it, dividing itself into the liquor, must diminish in some measure its fluidity, and consequently prevent the colouring particles which are suspended in it, being precipitated too quick, in a liquor which had not acquired a certain degree of thickness.

Notwithstanding this distributed throughout the liquor, as well from the bran as the madder, which also affords something glutinous, the colouring particles will subside if the liquor remains some days without being stirred; then the top of the liquor gives but a feeble tint to the body dipped in, and if a strong one is wanted, the mixture must be raked, and left to rest an hour or two, that the iron in the copperas, and the gross parts of the lime may fall to the bottom, which otherwise would mix with the true colouring particles, and prejudice their dye, by depositing on the body to be dyed a substance that would have but little adhesion, which in drying would become friable, and of which each minute part would occupy a space, where the true colouring particle could neither introduce nor deposite itself by an immediate contact on the subject.

Not to deviate from the method followed by the dyers, I boiled one part of grape-madder and one of bran, in 174 parts of water; this proportion of water is not necessary, more or less may be put, but I wanted to fill my vessel, which contained 512 parts. I passed this bever through a cloth and squeezed it putting this liquor, still hot, and which was of a blood-red, into the indigo liquor, observing the necessary precautions to prevent the breaking of the glass vessel. The whole was well stirred, and two hours after the liquor was green, and consequently fit for dying. It dyed cotton of a lasting blue, somewhat brighter than it was before the addition of the red of madder.

I shall now endeavour to find out the particular cause of the solidity of this colour; perhaps it may be the general cause of the tenacity of all the rest; for it appears already, from the experiments above related, that this tenacity depends on the choice of salts which are added to the decoctions of the colouring ingredients, when the same ingredients contain none in themselves. If from the consequences which shall result from the choice of these salts, of their nature, and of their properties, it be admitted (and it cannot be fairly denied) that they afford more or less tenuity in the homogeneous colouring parts of the dying ingredients, the whole theory of this art will be discovered, without having recourse to uncertain or contested causes.

One may easily conceive that the salts added to the indigo vats not only open the natural pores of the subject to be dyed, but also unfold the colouring atoms of the indigo.

In the other preparations of dyes (to be mentioned hereafter) the woollen stuffs are boiled in a solution of salts, which the dyers call preparation. In this preparation tartar and allum are generally used. In some hours the stuff is taken out, slightly squeezed, and kept damp for some days in a cool place, that the saline liquor which remains in it may still act and prepare it for the reception of the dye of these ingredients, in the decoction of which it is plunged to boil again. Without this preparation, experience shows that the colours will not be lasting, at least for the greatest part; for it must be owned that there are some ingredients which yield lasting colours, though the stuff has not previously undergone this preparation, because the ingredient contains in itself these salts.

It is therefore necessary, that the natural pores of the fibres of the wool should be enlarged and cleansed by the help of those salts, which are always somewhat corroding, and perhaps they open new pores for the reception of the colouring atoms contained in the ingredients. The boiling of this liquor drives in the atoms by repeated strokes. The pores already enlarged by these salts, are further dilated by the heat of the boiling water; they are afterwards contracted by the external cold when the dyed matter is taken out of the copper, when it is exposed to the external air, or when it is plunged into cold water. Thus the colouring atom is taken in, and detained in the pores or fissures of the dyed body, by the springiness of its fibres, which have contracted and restored themselves to their first state, and have re-assumed their primary stiffness upon being exposed to the cold.

If, besides this spring of the sides of the pore, it be supposed that these sides have been plaistered inwardly with a layer of the saline liquor, it will appear plainly that this is another means employed by art to detain the colouring atom; for this atom having entered into the pore, while the saline cement of the sides was yet in a state of solution, and consequently fluid; and this cement being afterwards congealed by the external cold, the atom is thereby detained; by the spring which has been mentioned, and by this saline cement, which by crystalization is become hard, forms a kind of mastic which is not easily removed.

If the coloured atom, (which is as small as the little eminence that appears at the entrance of the pore, and without which the subject would not appear dyed) be sufficiently protuberant to be exposed to more powerful shocks than the resistance of the sides of the cement that retains it, then the dye resulting from all these atoms sufficiently retained, will be extremely lasting, and in the rank of the good dye, provided the saline coat can neither be carried off by cold water, such as rain, nor calcined or reduced to powder by the rays of the sun; for every lasting colour, or colour belonging to the good dye, must withstand these two proofs. No other can reasonably be expected in stuffs designed for apparel or furniture.

I know but of two salts in chymistry, which, being once crystalized, can be moistened with cold water without dissolving; and there are few besides these that can remain several days exposed to the sun, without being reduced to a flower or white powder. These are tartar, either as taken from the wine vessels, or purified, and tartar of vitriol. The tartar of vitriol may be made by mixing a salt already alkalized, (or that may become such when the acid is drove out with a salt whose acid is vitriolic, as copperas and allum); this is easily effected if it be weaker than the acid of vitriol, and such is the acid of all essential salts extracted from vegetables.

In the process of the blue vat, which I tried in small, to discover the cause of its effects, copperas and potash, (which is a prepared alkali) are mixed together; as soon as these solutions are united, the alkali precipitates the iron of the copperas in form of powder almost black; the vitriolic acid of the copperas, divested of its metallic basis by its union with the alkali, forms a neutral salt, called *tartar of vitriol*, as when made with the salt of tartar and the vitriolic acid already separated from its basis; for all alkalis, from whatever vegetables they are extracted, are perfectly alike, provided they have been equally calcined.

More difficulties will occur with regard to the water for the preparation of other colours, such as reds and yellows. It may be denied that a tartar of vitriol can result from the mixture of allum and crude tartar boiled together ; yet the theory is the same, and I do not know that it can be otherwise conceived. The allum is a salt, consisting of the vitriolic acid united with an earth; by adding an alkali, the earth is immediately precipitated, and the tartar soon forms; but instead of this alkaline salt, allum is boiled with the crude tartar, which is the essential salt of wine, that is, a salt composed of the vinous acid, (which is more volatile than the vitriolic) and of oil, both concentrated in a small portion of earth.

This salt, as is known to chymists, becomes alkali by divesting it of its acid. Thus when the allum and crude tartar are boiled together, besides the impression which the fibres of the stuff to be dyed receive from the first of these salts, which is somewhat corrosive, the tartar is also purified, and by the addition of the earth, which is separated from the allum, (and which has near the same effect upon the tartar, as the earth of *Merviels*, which is used at Montpellier in manufacturing cream of tartar) it becomes clear and transparent. It may very probably happen that the vitriolic acid of the allum, driving out a part of the vegetable acid of the tartar, a tartar of vitriol may be formed as hard and transparent as the crystal of tartar. Admitting one or other of these suppositions, consequently there is in the open pores of the wool a saline cement which crystalizes as soon as the stuff which comes out of the dye is exposed to the cold air, which cannot be calcined by heat, nor is soluble in cold water. I could not avoid making this digression.

This theory is common to the indigo vat, where urine is used instead of water ; allum and crude tartar in the place of vitriol and pot-ash. This urine vat gives a lasting dye only when used hot, and then the wool must remain in an hour or two to take the dye equally. As soon as the vat is cold she strikes no more dye; the reason of this would be difficult to discover in an opaque metal vat, but in a glass vessel it is easily seen.

I let this little glass proof vat cool, and all the green colour, which was suspended in it while hot, precipitated little by little to the bottom; for then the tartar crystalizing itself, and reuniting in heavier masses than its moculas were during the heat of the liquor, and its solution, it sunk to the bottom of the vessel, and carried with it the colouring particles.

When I restored this liquor to its former degree of heat, after shaking it, and letting it settle a while, I dipped a piece of cloth, which I took out one hour after, with as lasting a dye as at the first; so that when this vat is used and fit to work, the tartar is to be kept in a state of solution, which cannot be done but by a pretty strong heat. The alkali of the urine greens it, the allum prepares the fibres of the wool, and the crystal of tartar secures the dye by cementing the colouring atoms deposited in the pores.

There still remains a difficulty with respect to the indigo vat, in which, neither vitriol, allum or tartar are used, but only pearlash or a fixed alkali in equal quantity with the indigo, and which is pretty briskly heated to dye the wool and stuffs. But before I enter into the cause of the solidity of its dye, which is equal to that of the other blue vats where the other salts already mentioned enter, I must examine into the nature of pearlash; it is a vegetable fixed alkali obtained from ashes and are the salts of lies calcined; potash is of the same nature, and from the same source, but the process is a little different in manufacturing it; it is not so mild and pure as the pearlash, it contains a much larger quantity of earth, and operates in the dye more quick and active. Some erroneously formed an idea that these alkalis were the lees of wine, and lost their acid substance by calcination, as Mr. Haigh, (dyer of Leeds,) observes on the nature of pearl-ashes; "which are the lees of wine dried and calcined: it is therefore an alkaline salt, of the nature of salt of tartar, but less pure as proceeding from the heaviest parts of the dregs of wine, and consequently the most earthy; besides, the alkali of the pearlash is never as homogeneous as the alkaline salt of tartar well calcined, and there are scarcely any pearlash not putrified, from which a considerable quantity of tartar of vitriol may not beobtained; it is even probable by an experiment which I have related, that it might at length be entirely converted into this neutral salt; the same may be said of potash, and of all other alkaline salts, whose basis are not that of the marine salt."

This is an error of Mr. Haigh, for pot and pearl-ashes have not the least connection with tartar or lees of wine, or tartar of vitriol, and it cannot be converted into a neutral salt.

Mr. Haigh and all others that form this opinion, are in an error, for the alkali of pot and pearl-ashes and lime, have not the least share of acid in them. Whatever qualities they may be possessed of in nature, are hidden from us till reduced to atoms, by the elementary heat. The pot and pearl-ashes are hidden in the plants or vegetable world, some vegetables possess more oleatic than others; the lime is hidden in the earth, or stones, and its alkali substance is not discovered till it is obtained by fire and air. The animal world are more or less alkali; for instance, the oyster shell. The alkali is found by the same element; however, the animal world are more or less acidous, as will be shown hereafter in manufacturing indigo and woad.

Of Borax.—The nature of borax is a neutral salt, used to correct the acid that arises from the vegetable substances of wheat bran and madder, (not diluted) by fermentation creates an acid with the alkali of a volatile, urinous substance; and likewise the indigo and woad, have a certain degree of acid, uncertain to determine, and if the acid should have the advantage of the alkali, and is not discovered soon enough, the dye is lost; so it requires borax or some neutral substance to correct the acid, and to act with them both, and it cannot be affected only by neutral salts, or a substitute of the same nature.

I must now give the reason why the indigo vat is green under the first surface of the liquor; why this liquor must be green that the blue dye may be lasting, and why the stuff that is taken green out of the liquor becomes blue as soon as it is aired. All these conditions being of necessity common to all indigo vats either cold or hot, the same explication will serve for them all.

1. The flurry which rises on the surface of the indigo liquor when it is fit dye is blue, and the under part of this scum is green; these two circumstances prove the perfect solution of the indigo, and that the alkaline salt is united to its colouring atoms since it greens them, for without they would remain blue.

2. These circumstances prove that there is also in the indigo a volatile urinous alkali, which the fixt alkali of the potash, or the alkaline earth of the lime displays, and which evaporates very

shortly after the exposition of this scum to the air. The existence of this urinous volatile appears plainly by the smell of the vat during the fermentation; when stirred, or when heated, the smell is sharp, and resembles that of stinking meat roasted.

3. In the preparation of the anil, in order to separate the fecula, a fermentation is continued to putrefaction. All rotten plants are urinous. This volatile urinous quality is produced by the intimate union of salts with the vegetable oil, or is owing to a prodigious quantity of insects falling on all sides of fermenting plants, and attracted by the smell exhaling from them, where they live, multiply, and die in them, and consequently deposit a number of dead bodies; therefore to this vegetable substance an animal one is united, whose salt is always an urinous volatile. This same urinous quality exists also in the woad, which is prepared after the same manner, viz. by fermentation and putrefaction, and which will be further explained in the narrative of its preparation.

4. And lastly, if indigo or woad be distilled in a retort, either alone, or (which is much better) with some fixed saline or earthy alkali added to it, a liquor will be obtained, which, by all chymical essays, produces the same effects as volatile spirits of urine.

Why does not this volatile urinous quality in the indigo cause it to appear green, since it must be equally distributed through all its parts? And why does indigo, being dissolved in plain boiling water, tinge it blue and not green? It is because this volatile urinous salt is not concreted that it requires another body more active than boiling water to drive it out of the particles surrounding it; and the solution of indigo is never perfected by water alone; whatever degree of heat is given, it is only diluted, and not dissolved in it. Indeed this decoction of indigo blues the stuffs that are dipped, but the blue is not equally laid on, and boiling water almost instantly discharges it. I shall endeavour to answer this by an example drawn from another subject.

Salt ammoniac, from which chymists extract the most penetrating volatile spirit, has not that quick urinous smell by dissolving and boiling it in water; either lime, or fixed alkaline salt, must be added to disengage the urinous volatile parts. In like manner, the indigo requires fixed saline, or earthy alkalis, to be exactly discomposed, that its volatile urinous salt may be discovered, and that its colouring atoms may be reduced probably to their elementary minuteness.

I now come to the second quality required. The liquor of the indigo vat must be green, that the dye may be lasting; for the indigo would not be exactly dissolved, if the alkali did not act upon it. Its solution not being as perfect as it ought to be, its dye would be neither equal nor lasting; but as soon as the alkaline salts act upon it, they must green it: for an alkali, mixed with the blue juice or tincture of any plant or flower, immediately turns it green, when equally distributed on all its colouring parts. But if by evaporation these same parts, coloured, or colouring, have re-united themselves into hard and compact masses, the alkali will not change their colour till it has penetrated, divided, and reduced them to their primary fineness. This is the case with indigo, whose fecula is the dry inspissated juice of the anil.

With respect to the last circumstance, which is that the stuff must be green on coming out of the liquor and become blue as soon as it is aired, without which, the blue would not be of a good dye, the following reasons may be given: it is taken out green because the liquor is green; if it was not, the alkaline salt put into the vat would not be equally distributed, or the indigo would not be exactly dissolved. If the alkali was not equally distributed, the liquor contained in the vat would not be equally saline: the bottom of this liquor would contain all the salt; the upper would be insipid. In this case the stuff dipped in would neither be prepared to receive the dye, nor to retain it; but when it is taken out green at the end of a quarter of an hour's dipping it is a proof that the liquor was equally saline, and equally loaded with colouring atoms; it is also a sign, that the alkaline salts have insinuated themselves into the pores of the fibres of the stuff and enlarged them, as has been observed, and perhaps have formed new ones. Now there can be no boubt that an alkaline salt may have this effect on a woollen stuff, when it is evident that a very sharp alkaline lie burns and dissolves almost in an instant a flock of wool or a feather.

A process in dying called, by the French, *fonte de bourre*, that is the melting or dissolving of flock or hair, is still a further example The hair, which is used and boiled in a solution of pearlash in urine, is so perfectly dissolved as not to leave the least fibre remaining. Therefore if a lixivium, extremely sharp, entirely destroys the wool, a lie which shall have but a quantityof alkaline salt sufficient to act on the wool without destroying it, will prepare the pores to receive and preserve the colouring atoms of the indigo.

The stuff is aired after being taken green out of the vat, and after wringing it becomes blue. What is done by airing? it is cooled; if it is the urinous volatile detached from the indigo which gave it this green colour, it evaporates, and the blue appears again; if it is the fixed alkaline that causes this green, not only the greatest part is carried off by the strong expression of the stuff, but what remains can have no more action on the colouring part, because the small atom of tartar of vitriol, which contains a coloured atom still less than itself, is crystalized the instant of its exposition to the cold air, and contracting this same colouring atom by the help of the spring at the sides of the pore, it entirely presses out the remainder of the alkali, which does not crystalize as a neutral salt.

The blue is roused, that is, it becomes brighter and finer by soaking the dyed stuff in warm water, for then the colouring particles, which had only a superficial adherence to the fibres of the wool are carried off. Soap is used as a proof of the lasting of the blue dye, and it must stand it, for the soap, which is only used in a small quantity in proportion to the water, and whose action on the dyed pattern is fixed to five minutes, is an alkali, mitigated by the oil, which cannot act upon a neutral salt. If it discharges the pattern of any part of its colour, it is because its parts were but superficially adhering; besides, the little saline crystal which is set in the pore, whose use is to cement the colouring atom, cannot be dissolved in so short a time, so as to come out of the pore with the atom it retains.

This treatise lays down the essay of a method of dying different from any hitherto offered. I appeal to philosophers, who would think little of a simple narrative of processes, if I did not at the same time give their theory. I shall follow this method in the other experiments on reds, the yellows, or other simple colours, as it is absolutely necessary to have a knowledge of them before entering on the compound, as these are generally but colours laid on one after the other, and seldom mixed together in the same liquor or decoction.

Thus having once the knowledge of what procures the tenacity of a simple colour, it will be more easily known, if the second colour can take place in the spaces the first have left empty without displacing the first.

This is the idea which I have formed to myself of the arrangement of different colours laid on the same stuff, for it appears to me a matter of great difficulty to conceive that the colouring atoms can place themselves the one on the other, and thus form kinds of pyramids, each still preserving their colour, so that from a mixture of the whole a compound colour shall result, and which, notwithstanding, shall appear uniform, and as it were homogeneous. To adopt this system, we must suppose a transparency in these atoms, which it would be difficult to demonstrate; and further, that a yellow atom must place itself immediately on a blue one, already set in the pore of the fibre of a stuff, and that it must remain there strongly bound, so that they must touch each other with extreme smooth surfaces, and so with every new colour laid on.

It is not easy to conceive all this, and it appears more probable, that the first colour has only taken up the pores that it found open by the first preparation of the fibres of the stuff; that on the side of these pores there remains more still to be filled, or at least spaces not occupied, where new pores may be opened to lodge the new atoms of a second colour, by the means of a second preparation of water, composed of corroding salts, which being the same as those of the first preparing liquor, will not destroy the first saline crystals introduced into the first pores.

What has been already said with regard to the

indigo vat, may also serve to explain the action of the woad vat on wool and stuffs ; it is only supposing in the woad, that salts do naturally exist, pretty near of affinity to those that are added to the indigo vat. It appears by the description given of these vats, that the woad vat is by much the most difficult to conduct. I am convinced that these difficulties might be removed, if an attempt was made to prepare the isatis as the anil is in the West Indies. I shall therefore compare their different preparations. I have taken the following narrative from the memoirs of Mr. Astruc's *Histoire Naturelle du Languedoc*. *Paris*, *Cavalier* 1737, in 4to, p. 330 and 331.

"According to the opinion of dyers, woad only gives feeble and languishing colours ; whereas those of the indigo are lively and bright. This opinion I grant is conformable to reason : the indigo is a fine subtle powder ; consequently capable to penetrate the stuffs easily, and give them a shining colour. The woad, on the contrary, is only a gross plant, loaded with many earthy parts, which slacken the action and motion of the finer parts, and prevent them from acting effectually.

"I know but one way to remove this inconveniency, that is, to prepare the woad after the same manner the indigo is prepared ; by this means the colours obtained from the woad would acquire the lively and bright qualities of those procured from the indigo, without diminishing in the least the excellency of the colours produced by the woad.

"I have already made in small* experiments

* As this ingenious man has succeeded in small experiments, it is probable he would also in the large ones ; and then this plant easily cultivated in America would well recompence the pains of the husbandman.

on what I propose, and those experiments have succeeded, not only in the preparation of the powder of woad, but also in the use of this powder for dying."

It is incumbent on those who have the public good at heart, to cause trials at large to be made, and if they have the success that can reasonably be expected, it will be proper to encourage those who cultivate woad, to follow this new method of preparing it, and offer premiums to enable them to sustain the expenses this new pactice will engage them in, until the advantage they will reap from it may be sufficient to determine them to follow it.

I shall now propose the means to succeed in Mr. Astruc's experiments, and these means naturally result from considering the method used in Languedoc for the preparation of woad, and the ingenious method by which they separate the fecula of the anil in America. I shall give the preparation of this last in the sequel; those who desire a fuller description may consult *l'Histoire des Antiles du P. du Tertre & du P. Labat.* The following preparation of the pastel, or garden woad, is thus described by Mr. Astruc.

The preparation of indigo and potash.—The preparation of potash requires no other performance than to dissolve it in warm water, with constant stirring; say one gallon of water to every two pounds of potash, let it stand and cool, and keep it from filth and dirt, be careful and not have it disturbed, that the earthly parts may settle to the bottom, and the lie poured off by inclination, leaving the lees to be cast away; the pot and pearl-ashes must be kept in a clean tight vessel, to exclude it from the air, otherwise it will dissolve and loose its substance, and you cannot ascertain its qualities.

For indigo.—All indigo requires to be pulverized to a powder, or ground to a paste, let it be used in what dye it will, but for blue I shall give the several processes that I conceive to be the most correct. In the first place,.take and weigh the quantity of indigo required for setting or recruiting your dye; then wash it with clean water, pour off the water, it will take all the loose dirt; then beat it small that the balls or grinding may be performed, then take as much of the potash lie, prepared as above, as is necessary to have the balls run free, and the grinding done with ease, grind it to a paste; or if this is neglected and it is not ground to a paste or powder, the indigo is lost, for it will not dissolve in the dye, as some erroneously imagine, but becomes coated and congealed, and looses its active part with the other ingredients. This is the preparation of indigo for the blue vat, let it always be ready ground before setting your vat; set it aside, covered close to prevent evaporation, and to keep the dirt and filth from it. Some indigo will be differently prepared, or with different alkalis, but the grinding must be the same. Lime waters: (after the preparation of the lime, which will be given hereafter,) when it is necessary to use the lie of lime, take two quarts of lime to every gallon of water, put them in a tub, stir them well together, let stand twelve hours; then pour off the lie to your indigo to your liking; some processes will be to use sig or urine, when this is necessary take one bushel of ashes, one peck of stone lime, put them in a leech, wet them with warm water, then leech as much sig till the strength is out of your leech, wet your indigo, &c.; to dissolve the indigo with vinegar a vegetable acid, the indigo is placed with the vinegar in a kettle over a moderate fire and kept warm twenty-four hours, that the acid may evaporate, &c.

Preparation of Lime.

That the lime may be properly slacked for the dyer's use, take some convenient place to pour water on the lime till it begins to slack and crack, then cast it into an empty vessel, where the lime finishes slacking, and reduces itself to powder, considerably augmenting its bulk; it is afterwards sifted through a canvas, and kept in a dry hogshead.

Sour liquors are not only necessary in some circumstances of setting a woad vat, but also in some of the preparations given to wool and stuffs previous to their being dyed; they are prepared after the following manner:

Preparation of sour Liquors.

A copper of the size required is filled with river water, and when it boils, it is flung into a hogshead, where a sufficient quantity of bran has been put, and stirred with a stick three or four times a day. The proportion of bran and water is not very material; I have made a good liquor by putting three bushels of bran into a vessel containing seventy gallons. Four or five days after, this water becomes sour, and consequently fit for use in all cases, where it will not be detrimental to the preparations of wool that are independent of dying.

For it may happen, that wool in the fleece which has been dyed in a liquor where too great a quantity of sour water has been put, will be harder to spin, as the sediment of the bran forms a sort of starch that glues the fibres of the wool, and prevents them from forming an even thread. I must here take notice of the bad custom of letting sour liquors remain in copper-vessels, as I have seen in some eminent dye-houses; for this liquor being an acid, corrodes the copper, and if it remains long enough to take in a portion of this metal, it will cause a defect both in the dye and in the quality of the stuff: in the dye, because the dissolved copper gives a greenish cast; in the quality of the stuff, because the copper dissolved preys on all animal substances. The dyers are often ignorant of the cause of these defects.

I flatter myself I shall omit no essential point on the woad vat: if any difficulties or accidents, which I have mentioned, are not found in the practice they are not considerable, and an easy remedy will be found by those who make themselves familiar with the working part.

The readers who have no idea of this work, may think me too prolix, and find repetitions; but those who intend to make use of what I have taught in this chapter, will perhaps reproach me for not having said enough on the subject.

Those that read this chapter with attention, will not be surprised that the master-piece for apprentices to dyers of the great dye, is, to set the woad vat and work her.

Receipt 120*th. To set a vat of* 24 *barrels, as practised in America.*

Take 12lb. of potash, dissolve as before described; 16lb. of good indigo prepared and ground as before directed, (or if you have woad omit 4lb. of indigo,) and add 16lb. of woad, take 16lb of madder and 16 quarts of wheat bran, and weigh 3-4 of a pound of borax.

The setting.—To cleanse the water, take about twelve bushels of ashes with a half bushel of stone lime and let all the water run through this leech to cleanse it for your blue, when the water is thus prepared, fill the vat with it scalding hot; then fill your boiler with the leeched water, then add the madder, wheat bran, and half the potash lie that remains after grinding the indigo already prepared; heat this near boiling hot with constant stirring, then empty this in the vat by a spout, with the vat covered close; then fill the boiler as before, put in the indigo and the remaining potash lie, leaving the sediment behind, then the woad, (if you have any); all to be added when cold, heat moderately, with constant stirring till it boils, empty it in the vat, fill the vat to within twelve inches of the top, rake well, cover close, and let stand three hours; then add the borax, rake well and let stand ten hours; then have all ready prepared; if necessary and the dye has not come to work, have lime water prepared as before described, to four gallons water, eight quarts of lime; add one gallon of the lie, rake well and add of the lime lie every three hours, till the lime water is used; if it does not come to work, have another liquor prepared, take two bushels of ashes, and one peck of lime, wet with warm water, then leech through ten gallons of sig; feed the dye with this when you rake, till it comes to work, observing to keep the vat covered close to let the heat be kept regular and not too low, if it cools too much keep a small fire in the flew. Another sure remedy, have a few gallons of good lively malt, and plenty of hops-beer in fermentation fit for drinking, add this if necessary, if the dye does not come to work in time, forty-eight hours, rake well, (or you may add a pound or two of pearlash and rake well.)

To know when a dye has come to work.

A vat is fit to work when the grounds are of a green brown, when it changes, on its being taken out of the vat, when the flurry is of a fine Turkish or deep blue, and when the pattern, which has been dipt in it for an hour, comes out of a fine deep grass green. When she is fit to work, the bever has a good appearance, clear and reddish, and the drops and edges that are formed under the rake in lifting up the bever are brown. Examining the appearance of the bever, is lifting up the liquor with the hand or rake, to see what colour the liquor of the vat has under its surface. The sediment or grounds must change colour (as has been already observed) at being taken out of the bever, and must grow brown by being exposed to the external air. The bever or liquor must feel neither too rough nor too greasy, and must not smell either of lime or lie. These are the distinguishing marks of a vat that is fit to work.

Wool and woollen stuffs of all kinds, are dyed blue without any other preparation then wetting them well in luke-warm water, with the addition of pearlash as before described, squeezing them well afterwards, or letting them drain: this precaution is necessary, that the colour may the more easily insinuate itself into the body of the wool, that it may be equally dispersed throughout; nor is this to be omitted in any kind of colours, whether the subject be wool or cloth.

When the vat has come to work, it must stand one hour after raking, then open it, take off the flurry or head with your skimmer, and put it in a tub and cover close, that it may be returned into the vat again, when you cover and rake, after dipping your goods.

The vat being come to work, the cross must be let down, and about thirty ells of cloth, or the equivalent of its weight of wool well scoured, (which is first intended to be dyed of a Persian blue to make a black afterwards), having returned this stirring several times, which must have always been covered with liquor, the cloth must be twisted on the rings fastened to the jack at the top of the vat; if it be wool, it is to be dipt with a net, which will serve to wring it: the cloth must be opened by its lists to air it, and to cool the green, that is, to make it lose the green colour it had coming out of the vat, and take the blue.

In the preceding work, I gave particular directions for the utensils, and the management of the cloths while in the vat; the same processes are to be observed in the management of the blue for cloths; wool is placed in a net and kept loose and open by poles for that purpose, that a man may raise the wool and loosen it by keeping one end of the pole in his hand, and the other in the dye with constant stirring, raising the wool but not exposed to the air, till taken up by the net for that purpose, for if the air turns it from the green to blue before it comes out of the vat, it will cause it to take the dye uneven, caution must be used not to crowd the dye too fast, and never to keep the vat long open, not to exceed three hours at a time, before you return your head, cover close and rake well; if the dye does not colour quick or active enough, add when your cover and rake; one pound of pearlash or more according to the state of the dye, judgment must be used; be ever mindful to keep the heat regular, if it gets too low, it will retard business, and you must let it stand some hours, it is not good to have it over hot, the dye will not turn to as much profit; keep it near to scalding heat, these kind of vats are very easily managed with attention, as the dye does not require shifting to reheat as the other vats, till the dye is worked off; no additions are to be made unless the dye works too slow, then you may add pearlash or sig leech, and some madder if necessary, and wheat bran. When the dye has lost its colour, recruit in manner and form as in setting; if the dye grows thick, dirty and glutinous by use, dip off the top of the dye carefully and let the sediment be cast away, and the dye boiled and skimmed. These directions are to be general in all blue dying, except otherwise directed; almost every blue dyer pretends to a peculiar skill or secret in blue dying, and yet the principle is the same, for the colouring substances indigo and woad, we all depend upon, and the power that operates them, the alkali, pot and pearl-ashes, lime, ashes, sig, &c. All the difference is the changing the order of them, and applying the assisting subjects, as madder and wheat bran; I shall make my observations general under this, both for indigo and woad, in the management of the cloth, wool, and vat; then show the different methods in practice. To return to the vat;

If the vat be in good order at the first opening, three or four stirrings or dippings may be made, and the next day, two or three more, only observing not to hurry her, or to work her as strong as at first. That the vat may turn to as much profit as possible for the shades of blue; first, all stuffs intended to be black, are dyed; then the king's blue; after these the green brown: the violets and Turkish blues are commonly done in the last rakings of the second day of the opening. The third day, if the vat appears much diminished, she must be filled with hot water within four inches of the brim. This is called filling the vat.

The latter end of the week, the light blues are made, and on Saturday night, having raked the vat, she is to be served a little more than the preceding day, that she may keep till Monday.

Monday morning the bever is put on the fire, by passing it from the vat into the copper by a trough, which rests on both; this clear bever is emptied to the grounds, and when it is ready to boil it must be returned into the vat, raking the grounds, as the hot liquor falls from the trough; at the same time have your indigo prepared, and the same process is to be observed as in setting; it generally comes to work much sooner, (in about fourteen hours); manage in manner and form as before described, till you obtain the colour and shade required.

The Woad or Pastel vat; how managed and how to know when a Vat is cracked by too great or too small a quantity of Lime; extremes which must be avoided.

When more lime has been put in than was sufficient for the woad, it is easily perceived by dipping in a pattern, which instead of turning to a beautiful grass green, is only daubed with a steely green. The grounds do not change, the vat gives scarcely any flurry, and the bever has a strong odour of quick lime, or its lees.

This error is rectified by thinning the vat, in which the dyers differ; some use tartar, others bran, of which they throw a bushel into the vat, more or less in proportion to the quantity of lime used, others a pail of urine. In some places a large iron chafing-dish is made use of, long enough to reach from the ground to the top of the vat, this chafing dish or furnace has a grate at a foot distance from its bottom, and a funnel coming from under this grate, and ascending to the top of the chafing-dish, which is to give air to, and kindle the coals which are placed on the grate. This furnace is sunk in the vat, near to the surface of the grounds, so as not to touch them, and is fastened with iron bars to prevent its rising. By this method the lime is raised to the surface of the liquor, which gives an opportunity to take off with a sieve what is thought superfluous; but when this is taken out, the necessary quantity of ware must be carefully restored to the vat. Others again thin the vat with pearlash, or tartar boiled in stale urine: but the best cure, when she is too hard, is, to put in bran and madder at discretion; and if she be but a little too hard, it will suffice to let her remain quiet four, five, or six hours, or more, putting in only two quarts of bran and three or four pounds of madder, which are to be lightly strewed on the vat, after which it is to be covered. Four or five hours after, she is to be raked and plunged, and according to the colour, that the flurry which arises from this motion, assumes and imprints on the whole liquor, a fresh proof is made by putting in a pattern.

If she is cracked, and casts blue only when she is cold, she must be left undisturbed, sometimes whole days without raking; when she begins to strike a tolerable pattern, her liquor must be reheated or warmed; then commonly the lime, which seemed to have lost all power to excite a fermentation, acquires new strength, and prevents the vat from yielding its dye so soon. If she is to be hastened, some bran and madder are to be thrown on, as also one or two baskets of new woad, which helps the liquor that has been reheated to spend its lime.

Care must be taken to put patterns in each hour, in order to judge, by the green colour which they acquire, how the lime is worked on. By these trials she may be conducted with more exactness, for when once a vat is cracked, by too great or too small a quantity of lime, she is brought to bear with much more difficulty.

If while you are endeavouring to bring her to work, the bever grows a little too cold, it must be heated by taking off some of the clear, and instead therof, adding some warm water; for when the bever is cold, the woad spends little or no lime; when it is too hot, it retards the action of the woad, and prevents it from spending the lime; therefore it is better to wait a little, than to hasten the vats to come to work when they are cracked. A vat is known not to have been sufficiently served with lime, and that she is cracked, when the bever gives no flurry, but instead thereof gives only a scum, and when she is plunged or raked, she only works, ferments and hisses, (this noise is made by a great number of air bubbles that burst as soon as they form), the liquor has also the smell of a common sewer or sink, or rotten eggs; it is harsh and dry to the touch; the grounds when taken out do not change, which generally happens when a vat is cracked for want of lime. This accident is chiefly to be apprehended when a vat is opened and a dip made in her; for if her state has not been looked into, both in regard to the smell as well as raking and plunging, and that the stuffs be imprudently put in when the woad has spent its lime it is to be feared the vat may be lost; for the stuffs being put in, the small quantity of lime that still remains in a state to act, sticks to them, the bever is divested of it, and the stuffs only blotted; these must be immediately taken out, and a quick remedy applied to the vat, to preserve the remaining part of the dye, which is done by putting in three or four measures of lime, more or less, according as the vat is cracked, and that without raking her bottom.

It is also to be observed, that if in raking and plunging the fermentation ceases, and the bad smell change, it is then to be supposed that the.

bever or liquor alone has suffered, and that the grounds are not yet in want. When the fermentation is in part or totally abated, and the bever has a smell of lime, and feels soft to the touch, the vat is to be covered and left at rest; and if the flurry still remains on the vat an hour and a half, a pattern is to be put in which must be taken out one hour after, and you are to be guided according to the green ground it will take. But generally vats that are thus cracked, are not so soon brought to a state fit for dying.

I shall make same reflections necessary to attend a more perfect knowledge of this process. The woad vat must never be re-heated but when fit for working; that is, she must have neither too much nor too little lime, but be in such a state as only to want heating to come to work. It is known she has too much lime (as has been before observed) by the quick smell; on the contrary, a want is known by the sweetish smell, and by the scum which rises on the surface by raking, being of a pale blue; but when this woad vat has come to work the same process is to be observed as in the preceding, dip and air to give it the blue.

If the cloth or wool was not deep enough for a mazarine blue by the first dipping, it must get another, by returning into the vat the end of the piece of cloth which first came out; and according to the strength of the woad, you must give to this striking two or three returns, as may be thought necessary for the intensity of the blue required. If the woad be good, such as the true L'Auragais is commonly, after taking out the first stirring, a second may be put in at this first opening of the vat. After making this opening, which is also called the first raking, the vat is to be again raked, and served with lime at discretion, observing that it has the smell and touch conformable to what has been laid down before, and taking notice, that in proportion as the dye diminishes, so does the strength of the woad.

As has been observed, the latter end of the week the light blues are made, and on Monday morning the bever or dye liquor is put to boil as before described, and a kettle of indigo put in.

When the vat is filled within four inches of the brim, and well raked, she must be covered, and two hours after a pattern put in, which must remain not more than an hour ; lime must be added according to the shade of the green, which this proof pattern shall have taken, and at the expiration of an hour or two, if the vat has not suffered, the stuff is to be put in ; having conducted it between two waters for about half an hour it is wrung, and a dip is again given to it, as was done in the new vat. This vat heated again, is conducted in the same manner, that is, three rakings are made the first day, observing at each raking, whether she wants lime ; for in this case, the quantity judged necessary must be given.

Blue made of woad alone, according to the opinion of some persons prejudiced in favour of old customs, is much better than that which the woad gives with the addition of indigo. But then this blue would be much dearer, because woad gives much less dye than indigo, and it has been found by repeated experience, that four pounds of fine indigo from Guatimala, produced as much as a bale of Albigeois woad or pastel ; and five pounds as much as a bale from L'Auragais, which generally weighs two hundred and ten pounds. So the using of the indigo with the woad is a great saving, as one vat with indigo shall dye as much as three without it.

Indigo is generally put into new vats after the

woad yields its blue, and a quarter or half after she is to be served with lime; as this solution of indigo is already impregnated with some of its dissolution, the lime must be given with a more sparing hand than where the woad is used alone. At the re-heating, the indigo is put in on Saturday night, that it may incorporate with the bever, and that it may serve as garnish by its lime. The indigo that is brought from Guatimala in America is the best; it is brought over in the shape of small stones, and of a deep blue; it must be of a deep violet colour within and when rubbed on the nail, have a copper hue; the lightest is the best. It is necessary to observe that for the better conducting of a woad vat, and to prevent accidents, a manufacturer ought to have a good woadman, this is the name given to the journeyman dyer, whose principal business is to conduct the woad; practice has taught him more than this treatise can furnish.

Care must be taken when a vat is intended to be re-heated, not to serve her with lime in the evening, (unless in great want of it) for if she was too much served with it, she might next day be too hard, as the dyers term it; for by heating her again, a greater action is given to the lime, and makes her spend it the quicker. Fresh indigo is commonly put into the vat, each time she is re-heated, in proportion to the quantity to be dyed. It would be needless to put in any, if there was but little work to do, or only light colours wanted. It was not permitted by the ancient regulations of France, to put more than six pounds of indigo to each bale of woad, because the colour of the indigo was thought not lasting, and that it was only the great quantity of woad which could secure and render it good; but it is now ascertained, both by the experiments of Monsieur Dufay, and those which I have since made, that the colour of indigo, even used alone, is full as good, and resists as much the action of the air, sun and rain, as that of pastel or woad.

When a vat has been heated two or three times, and a good part has been worked off, the same liquor is often preserved, but part of the grounds are taken out, which is replaced by new woad; (this is called vamping); the quantity cannot be prescribed on this occasion, for it depends upon the work the dyer has to do. Practice will teach all that can be wished for on this head. There are dyers who preserve liquor in their vats several years, renewing them with woad and indigo in proportion as they work them; others empty the vat entirely, and change the liquor when the vat has been heated six or seven times, and that she gives no more dye. A series of practice alone will show which of these is preferable. It is however more reasonable to think, that by renewing it now and then, more lively and beautiful colours may be obtained, and the best dyers follow this method.

In Holland they have vats which do not require to be so often heated. Mr. Van Robbais had some of these made some years since for their royal manufactory at Abbeville. The upper parts of these vats, to the height of three feet, are of copper, and the rest lead. They are also surrounded with a small brick wall, at seven or eight inches from the copper; in this interval embers are put, which keep up the heat of the vat a long time, so that she remains several days together in a condition to be worked, without the trouble of heating her over again. These vats are much more costly than the others, but they are very convenient, especially for the dipping of very light colours; because the vat is always fit to work, though she be very weak; this is not the case of the others, which generally make the colour a great deal deeper than required, unless they are set to cool considerably, and then it happens that the colour is not so good, nor has it the same brightness. To make these light colours in common vats, it is better to work some purposely that are strong with woad and weak of indigo; such give their colours slower, and light colours are made with greater ease.

As to the vats made in the Dutch fashion, and which have already been mentioned, the four which Mr. Van Robbais has in his manufactory, are six feet in depth, of which three feet and a half in the upper part are copper, and the two feet and a half of the bottom are lead. The diameter at the bottom is four feet and a half, and that at the top five feet four inches.

To return to the observations on heating the common vats. If the vat was heated when cracked, that is, when she has not quite lime enough, she would turn in the heating without being perceived, and perchance be entirely lost as the heat would soon finish the spending of the lime, which was in too small a quantity. If this is perceived in time, it must be helped by pouring it back into the vat without more heating; then feed her with lime, and not heat her till she is come to work.

On the re-heating, some of the grounds must be put into the copper with the liquor or bever; and great care must be taken not to boil it because the volatile necessary in this operation would evaporate. There are some dyers, who, in heating their vats, do not put the indigo immediately after the liquor is poured from the copper into the vat, but wait some hours till they see her come to work: this they do as a precaution, lest the vat should fail, and the indigo be lost; but by this method, the indigo does not so freely yield its colour, as they are obliged to work her as soon as she is fit, that she may not cool, so that the indigo, not being entirely dissolved, nor altogether incorporated, has no effect. It is therefore better to put it into the vat at the same time the liquor is cast in, and rake her well after. If the vat is heated over again without her coming to work, she must not be scummed as in the common heatings as the indigo would be carried off thereby, whereas, when she has worked, this scum is formed of the earthy part of the indigo and woad, united with a portion of lime.

When too much lime is put into a vat, you must wait for her till such time as she has spent it, or it may be accelerated by heating it, or by putting in ingredients which destroy in part the action of the lime, such as tartar, vinegar, honey, bran, some mineral acid, or any matter that will become sour; but all these correctors wear out the dye of the indigo and woad, so that the best method is, to let it spend of its own accord. A vat is not commonly fed with lime, but on the first, second, and sometimes the third day, and it is also remarked, not to dip the violets, purples, or any other wool or stuffs which have previously a colour that may be easily damaged; the succeeding day after its being fed with lime, as it is then too active, it dulls the first colour; the fifth or sixth day the crimsons may be dipt to give them a violet, and the yellows for green; following this rule, the colours will always be bright.

When a vat has been re-heated, she must come to work before she is served with lime; if this was done a little too soon, she must be cracked; the same thing would happen if some of the grounds were put into the copper. The most effectual method in this case is to let her rest before she is worked, until she comes to, which often happens in two, three, or four hours, and sometimes a day. By using light or weak lime, she grows too hard; because this light lime remains in the liquor, and does not incorporate with the grounds. This is known by the strong smell of the liquor, and on the contrary the grounds have a sweetish smell, whereas the smell ought to be equal in both. The best way then is, to let it spend itself, by raking her often, in order to mix the lime with the grounds, until the smell of the vat is restored, and the flurry becomes blue.

A woad vat may be set without the addition of indigo, but then she yields but little colour, and only dyes a small quantity of wool or stuffs; for one pound of indigo, as has already been observed, affords as much dye as fifteen or sixteen pounds of woad. I set one of this kind to try the qualities of woad by itself, and I could not find that indigo was any way inferior to it, either for the beauty or solidity of the colour. As lime is always used, and sometimes sour liquors, in the setting of a woad vat their preparation are spoken of in the preceding.

Receipt 121*st*. *Another method for blue, as practised in America.*

To set a vat of nine barrels, fill your vat about half full of boiling water, put two pounds of potash dissolved as before described, then add twelve quarts of wheat bran clear from the kernel, sprinkle it into the vat with your hand, then take one pound of good madder, then with the rake mix it with your dye, then add two pounds of indigo well ground, wet with urine, cover the vat closely; when you have introduced all the ingredients, the indigo being the last article, rake well and cover close, if possible to exclude the circulation of the air; let it remain eight or nine hours, then plunge and rake well with exertion and activity; bubbles will appear by repeating the plunges, and if a thick blue froth rises on the surface of the dye, which is called the head, continuing to float and the dye appears of a darkish green, the dye is in a good state, and is fit for colouring; it may be necessary to repeat the plunging and raking three or four times, remember after you have done raking to cover close; keep the heat regular. If the dye should cool before it comes to work you would have to reheat, but if you have a flew round your vat you may keep up the heat and save trouble; if the dye when opened in the morning appear of a pale blue cast, instead of a dark green, a handful or two of madder, say half a pound must be sprinkled in the vat; the dye should continue the heat near scalding. If the dye appear of a pale colour, two quarts of lime water must be added; be cautious not to open the vat often; let it stand at least two hours between each raking. After the vat is set and come to a head, let it stand secure till employed for dying: when the goods are ready for colouring, the dye must be heated, and add three pounds of indigo as before, together with the same proportions of potash, madder and wheat bran, and six pounds of woad, heat hot, and fill the vat within four inches of the top, cover close and follow the same processes in plunging and raking as before. If the dye is in good state there will be ten or twelve quarts of froth or head floating on the surface of the dye, the colour of which will be of a beautiful dark blue and the dye of a dark green: this is the proper state of the dye; have your goods prepared in hot water with pearlash, take it up, let drean, open the vat, take off the head and follow the same process; (if cloth, as in receipt No. 1; if wool, as in receipt No. 120), the utensils are the same in all blue dying of wool and woollen goods, the cloth when first taken out of the vat will exhibit a green appearance, by being exposed to the air will become blue, fold it over till well exposed to the air and all turned blue; be cautious and not expose any part of the goods to the air, to take off the green while in the vat, it will make the goods uneven; give your goods three or four stirrings or dippings, till your colour suits: put back the head, cover close and rake well, and let stand one hour, never dip till the sediment is well settled; when the liquor is thick and glutinous by use, it must be boiled, and the scum taken off and returned into the vat, add one gallon of lime water to cleanse the dye and settle the grounds. In hot weather if you are not using your dye, it must be heated as often as once in sixty days and raked frequently; when all your goods are dyed open the vat and give it the air till cold, then cover to keep out the insects, &c.

Receipt 122*d*. *Of setting and working a vat as practised at Paris, in France.*

It is a vat which is about five feet in height, two feet diameter, and becomes narrow towards the bottom; she is surrounded with a wall that leaves a space round her, which serves to hold embers. In a vat of this size, two pounds of indigo may at least be used, and five or six for the greatest proportion. To set a vat of two pounds of indigo in such a vessel that may contain about twenty gallons, about fifteen gallons of river water are set to boil in a copper for the space of half an hour, with two pounds of pearl-ash, two ounces of madder, and a handful of bran ; during this, the indigo is prepared after the following manner :

Two pounds of it are weighed out, and cast into a pail of cold water to separate the earthy parts. The water is afterwards poured off by inclination, and the indigo well ground ; a little warm water is put into it, shaking it from side to side ; it is poured by inclination into another vessel ; what remains is still ground, and fresh water put in to carry off the finest parts, and thus continued till all the indigo is reduced into a powder, fine enough to be raised by the water. This is all the preparation it undergoes. Then the liquor which has boiled in the copper with the grounds is poured into the high and narrow vat, as likewise the indigo ; the whole is then raked with a small rake, the vat is covered, and embers placed round her. If this work was begun in the afternoon, a few embers are added at night ; the same is repeated the next day morning and night. The vat is also lightly raked twice the second day ; the third day, the embers are continued to be put round, to keep up the heat of the vat ; she is raked twice in the day : about this time, a shining copper-coloured skin begins to apppear on the surface of the liquor, and appears as if it was broken or cracked in several places. The fourth day, by continuing the fire, this skin or pelicle is more formed and closer ; the flurry, which rises in raking the vat, appears, and the liquor becomes of a deep green.

When the liquor is in this state, it is a sign that it is time to fill the vat. For this purpose a fresh liquor is made, by putting into a copper about twenty quarts of water, with one pound of pearlash, a handful of bran, and half an ounce of madder. This is boiled a quarter of an hour, and the vat is served with it; she is then raked, and causes a great quantity of flurry to rise, and the vat comes to work the next day; this is known by the quantity of flurry with which she is covered by the skin or copper-scaly crust which swims on the liquor, which, although it appears of a blue-brown, is nevertheless green underneath.

This vat was much longer coming to its colour than the others, because the fire was too strong the second day, otherwise she would have been fit to work two days sooner. This did no other damage but retarded her, and the day she came to work, we dipt in serges weighing thirteen or fourteen pounds. As this caused her to lose her strength, and the liquor being diminished by the pieces of stuff that had been dyed in her, she was served in the afternoon with fresh liquor, made with one pound of pearlash, half an ounce of madder, and a handful of bran; the whole was boiled together in a copper for a quarter of an hour; the vat being served with it, she was raked, covered, and a few embers put round. She may be preserved after this manner several days, and when she is wanted to work, she must be raked over night, and a little fire place about her.

When there is occasion to re-heat, and add indigo to this kind of vat, two-thirds of the liquor (which then is no more green, but of a blue-brown and almost black) is put into a copper; when it is ready to boil, all the scum that is formed at the top is taken off with a sieve; it is afterwards made to boil, and two handfuls of bran, a quarter of a pound of madder, and two pounds of pearlash are added. The fire is then removed from the copper, and a little cold water cast into it to stop the boil; after which the whole is put into the vat, with one pound of powdered indigo, diluted in a portion of the liquor as before related ; after this the vat is raked, covered, and some fire put round ; the next day she is fit to work.

When the indigo vat has been re-heated several times, it is necessary to empty her entirely, and to set a fresh one, or she will not give a lively dye ; when she is too old and stale, the liquor is not of so fine a green as at first.

I put several other vats to work after the same method, with different quantities of indigo, from one pound to six ; always observing to augment or diminish the other ingredients in proportion, but always one pound of pearlash to each pound of indigo. I have since made other experiments, which proved to me that this proportion was not absolutely necessary ; and I make no doubt but that several other means might be found to make the indigo come to as perfect a colour. I shall, nevertheless, proceed to some other observations on this vat.

Of all those I set to work, after the manner described, one only failed me, and that by neglecting to put fire round her the second day. She never came to a proper colour ; powdered arsenic was put in to no effect ; red-hot bricks were also plunged in at different times ; the liquor turned of a greenish hue, but never came to the proper colour ; and having attempted several other means without success, or without being able to find out the cause of her not succeeding, I caused the liquor to be emptied and cast away.

All the other accidents that have happened me in conducting the indigo vat, have only lengthened the operation ; so that this process may be looked upon as very easy when compared to that of the woad vat. I have also made several experiments on both, in which my chief view was to shorten the time of the common preparation; but not meeting with the desired success, I shall not relate them.

The liquor of the indigo vat is not exactly like that of the woad; its surface is of a blue-brown, covered with coppery scales, and the under part of a beautiful green. The stuff or wool dyed in this is green when taken out, and becomes blue a moment after. We have already seen that the same happens to the stuff dyed in the woad vat; but it is remarkable that the liquor of the last is not green, and yet produces on the woad the same effect as the other. It must also be observed, that if the liquor of the indigo vat be removed out of the vessel in which it was contained, and if too long exposed to the air, it loses its green and all its quality, so that, although it gives a blue colour, that colour is not lasting.

Receipt 123*d*. *The Cold Vat with Urine.*

A VAT is also prepared with urine, which yields its colour cold, and is worked cold: for this purpose four pounds of indigo are powdered, which is to be digested on warm ashes twenty-four hours, in four quarts of vinegar; if it is not then well dissolved, it must be ground again with the liquor, and urine is to be added little by little, with half a pound of madder, which must be well diluted by stirring the liquor with a stick; when this preparation is made, it is poured into a vessel filled with 63 gallons of urine; it matters not whether it be fresh or stale; the whole is well stirred and raked together night and morning for eight days, or till the vat appears green at the surface when raked, or that she makes flurry as the common vat; she is then fit to work, without more trouble than previously raking her two or three hours before. This kind of vat is extremely convenient, for when once set to work, she remains good till she be entirely drawn, that is till the indigo has given all its colour ; thus she may be worked at all times, whereas the common vat must be prepared the day before.

This vat may at pleasure be made more or less considerable by augmenting or diminishing the ingredients in proportion to the indigo intended to be made use of; so that to each pound of indigo add a quart of vinegar, two ounces of madder, 15 or 18 gallons of urine. This vat comes sooner to work in summer that in winter, and may be brought sooner to work by warming some of the liquor without boiling, and returning it into the vat; this process is so simple that it is almost impossible to fail.

When the indigo is quite spent, and gives no more dye, the vat may be charged again without setting a new one. For this purpose, indigo must be dissolved in vinegar, adding madder in proportion to the indigo, pouring the whole into the vat, and raking her night and morning, and evening as at first, she will be as good as before; however she must not be charged this way above four or five times, for the ground of the madder and indigo would dull the liquor, and in consequence render the colour less bright. I did not try this method, and therefore do not answer for the success ; but here follows another with urine which gives a very lasting blue, and which I prepared.

Receipt 124*th*. *Hot Vat with Urine.*

A pound of indigo was steeped twenty-four hours in four quarts of clear urine, and when the urine became very blue it was run through a fine sieve into a pail, and the indigo which could not pass, and which remained in the sieve, was put with fore quarts of fresh urine; this was so continued till all the indigo had passed through the sieve with the urine; this lasted about two hours. At four in the afternoon three hogsheads of urine were put into the copper, and it was made as hot as could be without boiling. The urine cast up a thick scum, which was taken up with a broom and cast out of the copper. It was thus scummed at different times, till there only remained a white and light scum; the urine, by this means sufficiently purified and ready to boil, was poured into the wooden vat, and the indigo prepared as above, put in; the vat was then raked, the better to mix the indigo with the urine: soon after, a liquor was put into the vat, made of two quarts of urine, a pound of roach-allum, and a pound of red tartar. To make this liquor, the allum and tartar were first put into the mortar, and reduced to a fine powder, upon which the two quarts of urine were poured, and the whole rubbed together, till this mixture, which rose all of a sudden, ceased to ferment: it was then put into the vat, which was strongly raked; and being covered with its wooden cover, she was left in that state all night; the next morning the liquor was of a very green colour; this was a sign she was come to work, and that she might have been worked if thought proper, but nothing was dyed in her; for all that was done, was only, properly speaking, the first preparation of the vat, and the indigo which had been put in was only intended to feed the urine, so that to finish the preparation the vat was let to rest for two days, always covered, that she might cool the slower; than a second pound of indigo was prepared, ground with purified urine as before. About four in the afternoon all the liquor of the vat was put into the copper; it was heated as much as possible without boiling; some thick scum formed on it which was taken off, and the liquor being ready to boil was returned into the vat. At the same time the ground indigo was put in, with a liquor made as above of one pound of allum, one pound of tartar, and two quarts of urine, a fresh pound of madder was also added; then the vat was raked, well covered, and left so the whole night. The next morning she was come to work, the liquor being very hot, and of a very fine green, she was worked with wool in the fleece, of which thirty pounds were put into the vat. It was well extended and worked between the hands, that the liquor might the more easily soak into it; then it was left at rest for an hour or two, according as lighter or deeper blues are required.

All this time the vat was well covered, that it might the better retain its heat, for the hotter she is, the better she dyes, and when cold acts no more. When the wool came to the shade of the blue required, it was taken out of the vat in parcels, about the bigness of a man's head, twisted and wrung over the liquor as they were taken out, till from green, as they were coming out of the vat, they became blue. This change from green to blue is made in three or four minutes. These thirty pounds being thus dyed, and the green taken off, the vat was raked, and suffered to rest for two hours, being all that time well covered; then thirty pounds more were put in, which was well extended with the hands, the vat was covered, and in four or five hours this wool was dyed at the height or shade of the first thirty pounds; it was then taken out in heaps, and the green taken off as before. This done, the vat had still some little heat, but not sufficient to dye fresh wool; for when she has not a sufficient heat the colour she gives would neither be uniform nor lasting, so that it must be re-heated, and fresh indigo put in as before. This may be done as often as judged proper, for this vat does not spoil by age, provided, that whilst she is kept without working, a little air is let into her.

Re-heating of the Vat with Urine.

About four in the afternoon, the whole liquor of the vat was put into a copper, and a sufficient quantity of urine added to this liquor, to make up the deficiency that had been lost by evaporation during the preceding work. This filling commonly takes eight or nine pails of urine the liquor was then heated and scummed as before, and when ready to boil, returned into the vat with a pound of indigo, and the liquor above described, consisting of allum and tartar, of each one pound, madder one pound, and two quarts of urine. After raking the vat well, and covering her, she was left at rest the whole night.

The next day she came to work, and sixty pounds of wool were dyed in her at twice as before. It is after this manner all the re-heatings must be done the evening before the dying, and these re-heatings may extend to infinity, as the vat, once set, serves a long time.

I must here observe, that the greater the quantity of indigo put in at once is, the deeper the blue: thus instead of one pound, four, five, or six pounds may be put in together; nor is it necessary to augment the dose of allum, tartar, or madder, of which ingredients the liquor is composed but if the vessel hold more than three hogsheads, then the dose of these must be augmented in proportion. The vat I have mentioned held three, and was too small to dye at one time a sufficient quantity of wool to make a piece of cloth, viz. fifty or sixty pounds; for this purpose it would be necessary that the vat should contain at least six hogsheads, and from this a double advantage would arise. 1. All the wool will be dyed in three or four hours, whereas dying it at twice, it takes eight or ten hours. 2. At the end of three hours, in which time the wool would be dyed, taken out, and the green taken off, the vat being yet very hot; after raking and letting her rest a couple of hours the same wool might be returned into her, which would heighten the colour very much; for all wool that has been dyed, aired, and the green taken off, always takes a finer colour than new or white wool, which might remain twenty hours in the vat.

Great care must be taken to air and take off the green of the dyed parcels of wool that are taken out of the vat hastily, that the air may strike them equally, without which the blue colour will not be uniform throughout the wool.

There are manufacturers who say that cloths, whose wool has received this ground of blue with urine, cannot be perfectly scoured at the fulling mill, even at twice; others vouch the contrary, and I am of opinion the last speak the truth; yet, if the first are right, it might be suspected that the animal oil of the urine becoming resinous by drying on the wool, or by uniting with the oil with which the wool is moistened; for its other preparations more strongly resist the fuller's earth and soap, than a simple oil by expression. To remedy this, the wool ought to be well washed in a running water after it is dyed, twisted, aired, the green taken off, and cooled. Be it as it may, the woad vat will always be preferred in the great dye houses to those kinds of indigo vats made with urine or otherwise; and for this reason, that with a good woad vat, and an ingenious woadman, much more work is despatched than with all the other blue vats.

I have described the indigo vats in this treatise, not with a design to introduce them in the large manufactories, but to procure easy means to the dyers in small, and small manufactories, to whom I wish this work may be of as much advantage as to the others.

Receipt 125th. For blue vat, with garden-woad, or pastel-woad.

The garden-woad is a plant cultivated in many parts of Holland and France, and might be in America, to the great advantage of the husbandman; it is made up in bales generally weighing from one hundred and fifty pounds to two hundred; it resembles little clods of dried earth, interwoven with the fibres of plants; it is gathered at a proper season, and laid up to rot, and then made into small balls to dry. Several circumstances are to be observed in this preparation; on this you may see the regulations of Mons. Colbert on dyes; the best prepared comes from the diocese of Alby in France.

The Vat set to work.

A copper as near as possible to the vat is filled with water that has stood for some time, or, if such water is not at hand, a handful of dyer's woad or hay is added to the water, with eight pounds of crust of fat madder. If the old liquor from a vat that has been used in dying from madder can be procured, it will save the madder and produce a better effect.

The copper being filled, and the fire lighted about three in the morning, it must boil an hour and a quarter, (some dyers boil it from two hours and a half to three); it is then conveyed by a spout into the woad vat, in which has been previously put a peck of wheat bran. Whilst the boiling liquor is emptying into the vat, the balls of woad must be put one after another into the vat, that they may be the easier broken, raked and stirred; this is to be continued till all the hot liquor from the copper is run into the vat, which, when little more than half full, must be covered with cloths somewhat larger than its circumference, so that it may be covered as close as possible, and left in this state four hours. Then it must be aired, that is, uncovered to be raked, and fresh air let in it; and to each bale of woad, a good measure of ware flung in; this is a concealed name for lime that has been slacked. This measure is a kind of wooden shovel, which serves to measure the lime grossly; it is five inches broad and three inches and a half long, containing near a good handful; the lime being scattered in, and the vat well raked, it must be again covered, leaving a little space of about four fingers, open, to let in air. Four hours, after, she must be raked, without serving her with lime; the cover is then put on, leaving, as before, an opening for the air; in this manner she must be let to stand for two or three hours. Then she may be raked well again, if she is not yet come to work; that is, if she does not cast blue at her surface, and that she works or ferments still, which may be known by raking and plunging with the flat of the rake in the vat; being well raked, she is to remain still for one hour and a half more, carefully observing whether she casts blue. She is then to be served with water, and the quantity of indigo judged necessary is to be put in; it is commonly used in a liquid state, the full of a dye-house kettle for each bale of woad; the vat being filled within six finger-breadths of her brim, is to be raked and covered as before; an hour after filling her with water, she must be served with lime, viz. two measures of lime for each bale of woad, giving more or less according to the quality of the woad, and what may be judged it will spend or take of lime.

I hope the reader will excuse my plainness; this treatise being wrote for the dyer, I must speak the language he is used to; the philosopher will easily substitute proper terms, which perhaps the workman would not understand. There are kinds of woad readier prepared than others, so that general and precise rules cannot be given on this head. It must also be remarked, that the lime is not to be put into the vat till she has been well raked.

The vat being again covered, three hours after a pattern must be put in, and kept entirely covered for an hour; it is then taken out to judge if she be fit to work. If she is, the pattern must come out green, and on being exposed a minute to the air, acquire a blue colour. If the vat gives a good green to the pattern, she must be raked, served with one or two measures of lime, and covered.

Three hours after, she must be raked, and served with what lime may be judged necessary; she is then to be covered, and one hour and a half after, the vat being pitched or settled, a pattern is put in, which must remain an hour to see the effects of the woad. If the pattern is of a fine green, and that it turns to a deep blue in the air, another must be dipt in to be certain of the effect of the vat. If this pattern is deep enough in colour, let the vat be filled up with hot water, or if at hand, with old liquor of madder, and rake her well. Should the vat still want lime, serve her with such a quantity as you may judge sufficient by the smell and handling. This done, she must be again covered, and one hour after put in your stuffs, and make your overture. This is the term used for the first working of wool or stuffs in a new vat.

Receipt 126*th.* *To set a field Woad Vat.*

I HAVE but little to say on this woad vat, different from that which has been related of the pastel or garden woad. The woad is a plant cultivated in Normandy, and prepared after the same manner the garden woad is in Languedoc. The method of cultivating it may be seen in the French "General Instructions on Dyes," of the 28th of March, 1671, from the article 259 to 288, where it treats of the culture and preparation of the pastel and woad. The woad vat is set at work after the same manner as that of pastel; all the difference is that it has less strength and yields less dye. There follows a description of the woad vat, which I carried on in small, and in a bath heat, similar to that of the pastel in the foregoing chapter.

I placed in a copper a small vessel containing fifty quarts, and filled two-thirds with a liquor made of river water, one ounce of madder, and a little weld, putting in at the same time a good handful of wheat bran and five pounds of woad. The vat was well raked and covered; it was then five in the evening; it was again raked at seven, nine, twelve, two, and four o'clock; the woad was then working, that is, the vat was slowly coming to work, as I have already related of that of the pastel.

Pretty large air bubbles formed themselves,

but in a small quantity, and had scarcely any colour. She was then served with two ounces of lime and raked. At five o'clock a pattern was put in; which was taken out at six, raking her; this pattern began to have some colour; another was put in at seven, at eight she was raked, and the pattern came out pretty bright; an ounce of indigo was then put in; at nine another pattern, at ten she was raked, and one ounce of lime was added, because she began to have a sweetish smell; at eleven a pattern, at twelve she was raked; it was thus continued till five, then three ounces of indigo were put in, at six a pattern, at seven she was raked. It would then have been proper to have served her with water, as she was at that time perfectly come to work, the pattern that was taken out being very green, and turning of a bright blue. But besides that I was fatigued, having sat up the whole night, I chose rather to put her back to the next day, to see her effect by day-light; and for that purpose, I put one ounce of lime, which kept her up till nine in the morning: from time to time patterns were put in, the last that was taken out was very beautiful; she was served with a liquor composed of water, and a small handful of bran. She was raked and patterns put in from hour to hour; at five she was come to work; she was afterwards served with lime, and raked to preserve her till she was to be re-heated.

Some time after I set another with the woad alone without indigo, that I might be able to judge of the lasting of the dye of the woad, which upon trial, I found to be as good as the pastel or garden woad. Thus all the superiority the pastel has on the woad, is, that the latter yields less dye than the former.

The little varieties that may be observed in setting these different vats at work, prove, that there are many circumstances in these processes that are not absolutely necessary. It appears to me, that the only important point, and that to which the greatest attention is to be given, is, in the conducting the fermentation with care, and not to serve her with lime, but when judged necessary by the indications I have laid down. As to the indigo being put in at twice, or altogether, a little sooner or later, it appears very indifferent. The same may be said of the weld, which I made use of twice, and suppressed the two other times, and of pearlash, which I added in a small quantity in the small pastel vat, and suppressed in the woad vat. In short, I believe, and it appears to me to a demonstration, that the greatest regard is to be had to the proper distribution of the lime, throughout the whole course of the working of the vats, either to set them at work, or to re-heat them. I must also add, that when a woad vat is set to work, she cannot be too often inspected into to know her state; for if there are some that are backward (which is attributed to the weakness of the woad) there are also others that very quickly come to work. I have seen a middling one of seventy pounds of woad, poisoned; because the woad man neglected to inspect her as often as she required, and she had been two hours fit to work before he discovered it; the grounds were entirely come up to the surface of the liquor, and the whole had a very sour smell; it was not possible to bring her back, and they were obliged to fling her away, as she would in a short time have contracted a fœtid smell. The retarding of the action of the vat may also proceed from the temperature of the air; for the vat cools a great deal sooner in winter than in summer: therefore it becomes necessary to watch it attentively, though commonly they are fourteen or fifteen hours before they come to work.

To the dyer.—Blue and brown require no preparation, it is sufficient that the wool be well scoured, which will be noticed in its proper place; the wool is to be wet as already described for blue, it suffices to dip it in the vat, stirring it well, and letting it remain in the vat more or less time, according to the state of the dye and the ground of the colour wanted; many colours require a blue shade to be given to the wool. It is an easy matter to dye wool blue, when the vat is once prepared, but it is not so easy first to prepare the vat, which is the most difficult part of the dyer's art; for this reason I have given the most exact and extensive rules in my power, in this and the preceding work.

I have endeavoured to make my explanations general on the properties and effects of dye stuffs, and laid down the different processes of setting and managing the blue vat, both of woad and indigo, for woollen. The receipts for cotton and linen dye, will be noticed under their proper head, &c. It is the earnest wish of the author, if any attempt to set a blue vat, from this book, that they attend strictly to the rules and directions here laid down, and not let it be a momentary study but search to the bottom and find out the the principle actors in the dye, and rule the dye, and not let it rule you; upon this principle you may do yourself and country justice. I shall leave the subject of the blue on woollen goods, after giving the process of manufacturing woad and indigo.

The manufacturing of Pastel or Garden Woad, as practised in France.

Peasants of Abbigevois distinguish two kinds of woad seed: the one violet colour, the other yellow; they prefer the former, because the woad that shoots from it bears leaves that are smooth and polished, whereas those that spring from the yellow are hairy; this fills them with earth and dust, which makes the woad prepared from them of a worse quality. This woad is called *pastelbourg*, or *bourdaigne*.

The woad at first shoots five or six leaves out of the ground, which stand upright whilst green; they are a foot long, and six inches broad; they begin to ripen in June; they are known to be ripe by their falling down and growing yellow; they are then gathered, and the ground cleared from weeds, which is carefully repeated each crop.

If there has been rain, a second crop is obtained in July; rain or dry weather advances or retards it eight days. The third crop is at the latter end of August; a fourth the latter end of September; and the fifth and last about the tenth of November. This last crop is the most considerable, the interval being longer. The plant at this crop is cut at the root from whence the leaves spring. This woad is not good, and the last crop is forbid by the regulations. The woad is not to be gathered in foggy or rainy weather, but in serene weather, when the sun has been out some time.

At each crop the leaves are brought to the mill to be ground, and reduced to a fine paste; this is to be done speedily, for the leaves when left in a heat ferment, and soon rot with an intolerable stench. These mills are like the oil or bark-mills, that is, a mill stone turns round a perpendicular pivot in a circular grove or trough, pretty deep, in which the woad is ground.

The leaves thus mashed and reduced to a paste, are kept up in the galleries of the mill, or in the open air. After pressing the paste well with the hands and feet, it is beat down and made smooth with a shovel. This is called the woad piled.

An outward crust forms, which becomes blackish ; when it cracks, great care must be taken to close it again. Little worms will generate in these crevices and spoil it. The pile is opened in a fortnight, well worked between the hands, and the crust well mixed with the inside ; sometimes this crust requires to be beat with a mallet to knead it with the rest.

This paste is then made into small loaves or round balls, which according to the regulations, must weigh a pound and a quarter. These balls are well pressed in the making, and are then given to another, who kneads them again in a wooden dish, lengthens them at both ends, making them oval and smooth. Lastly, they are given to a third, who finishes them in a lesser bowl dish, by pressing and perfectly uniting them.

The pastel or woad thus prepared is called *Pastel en Cocaigne ;* whence arises the proverb, *Pais de Cocaigne ;* which signfies a rich country, because this country* where the woad grows, enriched itself formerly by the commerce of this drug.

These balls † are spread on hurdles, and exposed to the sun in fine weather ; in bad weather they are put at the top of the mill. The woad that has been exposed some hours to the sun, becomes black on the outside, whereas that which has been kept within doors is gen-

* *L' Abigeois & Lauragois*

† There is a place in India, the name I do not recollect, where the anil is prepared after the manner of the woad, and the Indigo comes from it in lumps, containing all the useless parts of this plant. It is very difficult to prepare a blue vat with it.

erally yellowish, particularly if the weather has been rainy. The merchants prefer the former; this makes little difference as to its use; it is in general always yellowish, as the peasants mostly work it in rainy weather when they cannot attend their rural employments.

In summer, these balls are commonly dry in fifteen or twenty days, whereas in autumn those of the last crop are long in drying.

The good balls when broke are of a violet colour within, and have an agreeable smell; whereas those that are of an earthy colour and a bad smell are not good; this proceeds from the gathering of the woad during the rain, when the leaves were filled with earth. Their goodness is also known by their weight, being light when they have taken too much air, or rotton by not having been sufficiently pressed.

Powder of Woad.

Of these balls well prepared, the powder of woad is to be made; for this purpose a hundred thousand at least are required. A distant barn or a warehouse must be procured, larger or smaller according to the quantity intended to be made. It must be paved with bricks and lined with the same, to the height of four or five feet; the walls would be better to be of stone to that height, yet often the walls are only coated with earth; this coat breaking off and mixing with the woad is a great prejudice to it. In this place the balls are reduced to a gross powder with large wooden mallets. This powder is heaped up to the height of four feet reserving a space to go round, and is moistened with water; that which is slimy* is best provided

* I can see no reason why slimy water, and yet to be clear, is preferred It appears to me that clear river water would

it be clear ; the woad thus moistened, ferments, heats, and emits a very thick stinking vapour.

It is stirred every day for twelve days, flinging it by shovels full from one side to the other, and moistening it every day during that time ; after which no more water is flung on, but only stirred every second day; then every third, fourth, and fifth ; it is then heaped up in the middle of the place, and looked at from time to time to air it in case it should heat. This is the pastel or garden woad powder fit for sale to the dyers.

Mr. Astruc, to prove that the sale of woad formerly enriched the higher Languedoc, quotes the following passage from a book entitled *Le. Marchand.*

" Formerly they transported from Toulouze to Bordeaux, by the river Garonne, each year a hundred thousand bales of woad which on the spot are worth at least fifteen livers a bale, which amounts to 1,500,000 livers : from whence proceeded the abundance of money and riches of that country." Castel in his *Memoirs de' l' Histoire du Languedoc*, in 1633, p. 49.

The comparing of these two methods of preparing the woad and indigo may be sufficient to a person of understanding, who might be appointed to try, by experiments, the possibility of extracting a fecula from the isatis of Languedoc like that of the anil. It is neither the dyer or manufacturer that ought be applied to for that purpose ; both would condemn the project as a novelty, and it would require many experiments, which in general they are not accustomed to.

be more secure ; with this they would avoid the inconveniences that must attend a standing water, always filled with filth ; or of a muddy water, which contains useless earth and which must make the dye uneven.

I could wish this experiment was tried in great, so that at least fifty pound of this fecula might be got, that several vats might be set in case the first should fail. Whoever does try it, should be very careful to describe all the circumstances of the process. Perhaps it might not succeed at the first crop of the leaves of the woad, because the heat in June is not sufficient, but probably he might meet with success in August.

If this succeeds, there are without doubt several other plants of the same quality as the isatis, and which yields a like fecula.

It is also probable that the dark green of several plants is composed of yellow and blue parts; if by fermentation the yellow could be destroyed the blue would remain. This is not a chimerical idea, and it is easy to prove that some use might be derived from such an experiment.

Of making Indigo in America.

INDIGO is the fecula of a plant named *nill* or *anil*; to make it, three vats are placed the one over the other, in form of a cascade; in the first, called the steeper, the plant is put in with its leaves, bark and flowers*, and filled with water; some time after, the whole ferments, the water grows intensely hot, thickens, and becomes of a blue colour bordering on the violet; the plant, according to the opinion of some, deposing all its salts, and according to others, all its substance. In this state, the cocks of the steeper are turned, and all the water let out stained with the colouring parts of the plant into the second, called the beater; because this wa-

* In the village of Sargussa, near the town of Amadabat, the Indians only use the leaves of the anil; they fling away the rest of the plant. The best indigo comes from thence.

ter is beat by a mill or a machine that has long sticks, to condense the substance of the indigo, and precipitate it to the bottom. By this means the water becomes clear and colourless, like common water; then the cocks are turned, that the water may run off from the surface of the blue sediment; after which, other cocks are turned that are at the bottom that all the fecula may fall into the third vat, called the reposer, for it is there the indigo remains to dry; it is then taken out to be made into cakes, &c. See, on this subject, *Histoire des Antilles, pare le Pere Labat.*

At Pondicherry, on the coast of Coromandel, there are two kinds of indigo, the one a great deal finer than the other; the best is seldom used but to lustre their silks, the inferior in dying. They augment in price according to their quality; there is some which cost from 15 pagodas the bar (which weighs 48 pounds) to 200 pagodas. The most beautiful is prepared nigh Agra. There is also a very good kind that comes from Masilupatan and Ayanon, where the East-India Company have a factory. At Chandernagor it is called nil when it is prepared and cut to pieces. The indigo of Java is the best of all; it is also the dearest, and consequently few dyers use it. Good indigo ought to be so light as to float on the water; the more it sinks, the more it may be suspected of being adulterated by a mixture of earth, cinders, or pounded slates. It must be of a deep blue, bordering on the violet, brillant, lively, and shining; it must be finer within, and appear of a shining hue. Its goodness is tried by dissolving it in a glass of water; if it be unmixed and well prepared, it will dissolve entirely; if sophisticated, the foreign matter will sink to the bottom. Another method of trying it is by burning; good indigo burns entirely away, and when adulterated, the mixture remains after the indigo is consumed.

Powdered indigo is much more subject to adulteration than that which is in cakes; for it is a difficult matter that sand, powdered slates, &c. should unite so as not to form together in different places layers of different matters; and, in this case, by breaking the lump indigo, it is easily discovered.

You will see by the manufacturing of the woad and indigo, that a portion of the animal creation (reptiles, and insects), live and die in it; this creates an acidous, alkaline, urinous and volatile substance, and is the reason why the vat requires to be covered close, to prevent the evaporation of the colouring substances.

CHAP. II.

Receipt 127th. On Yellow Dying.

YELLOW is one of the five material or primitive colours, and the subjects are many, of which I shall give a catalogue. Yellow is governed by the power of the acid. I shall not in this, point out any particular process for dying of cloth, as that has been described in my former work; see the receipts for yellow, in them it was for cloth only, (the wool differs from cloth,) to use the same proportion of preparation and dye stuff, for twelve pounds of wool you would for sixteen pounds of cloth. This is to be a general rule in all dying; the process for the management of wool when dying, has been described; it is to be put in a net, and stirred with poles, to keep the wool open, that it may receive the colour even, &c.

Of the five primary colours mentioned in the introduction, two of them require a preparation given by non-colouring ingredients, which by the acidity and fineness of their earth, dispose the pores of the wool to receive the colours: the yellow, the red, and the colours derived from them, must be so treated; black must have a preparation peculiar to itself.

Of drugs and woods for yellow.—They are the weld or wold, savory, green-wood, the yellow-wood and the finugrick; these are those tolerated by the regulations in the good dye; weld gives the brightest dye, green-wood and savory are the best for the wool to be made green, as they incline and border on the green, the three others give good yellows. The yellows are classed in three, the straw, the pale, and the lemon yellows.

To the five drugs already mentioned for yellow, may be added a number of the good dye; the bark and root of barbary shrub, the bark of the ash-tree, the dock root, the leaves of the almond, peach and pear-trees, assmart, and saffron flowers, may all be considered in the good dye. Those belonging to the false dye, are turmerick, which gives a beautiful yellow but soon changes; fustic gives a good colour, but soon turns brown, and is excellent in brown; roucou or racourt, the grains of Avignon, and onion leaves are much the same, to which may be added many others; in short all leaves, barks, and roots, which by chewing shew some little astriction, gives yellows of the good dye, more or less fine according to the time they are boiled, and in proportion to the tartar and allum used in the liquor. There is no colour that produces so great a variety of colouring substances as the yellow ; there is such a difference in the qualities of these subjects, there can be no regular system adopted, but must be applied as the colour requires. The dyer must use his judgment for the rule to direct his proportion for the dying subjects.

For dying yellow.—The common preparing water with tartar and allum, are used for wool or stuffs ; to each pound of wool take one ounce of tartar and four ounces of allum, or to every hundred weight of wool, twenty-five pounds of allum and six pounds of tartar ; put this into your copper caldron, fill with fair water, heat boiling hot, then immerse the wool, stir with poles to keep the wool open, that it may all receive the preparation alike ; boil six hours, take it up, let cool, place it in a sack, covered close, to lay twenty-four hours that the pores of the wool may inhale the salts, and be the better prepared ; then rince well and shift the liquor from your copper, clean well, fill with clean fair water ; if the waters are hard, or impregnated with minerals, to every hundred gallons of water, take four quarts of wheat bran, enclose it in a clean linen bag, let it boil one hour, or you may add three or four pails full of sour water ; map off the scum that rises by the heat. The hard and rough waters which are natural to some wells and places, by this process may be rendered soft and fit for any colour ; the cleansing of the waters requires strict attention in all light and bright colours, as the yellow, the red, &c. ; when the water is thus prepared, add of your colouring substances be they weld, yellow wood, roots, leaves or plants, they all require boiling ; add, boil and run, or stir, till you obtain the colour required.

Light shades of yellow are obtained in the same manner as all others spoken of, only the preparing liquor for these light yellows must be weaker. I recommend twelve pounds and a half of allum for each hundred pounds of wool, and the tartar in proportion ; but these light shades do not resist the proofs as deeper shades do, made with the full proportion of tartar.

Some dyers endeavouring to help this, leave the wool and stuffs for a longer time in the dye, because they take it slower in proportion to the weakness of the liquor : but if they put at the same time in the colouring liquor, wools whose preparation shall have been different, they shall take at the same time different shades. These liquors more or less strong are called half-preparing liquors, or quarter-preparing liquors, and they make great use of them in light shades of wool dyed in the fleece, that is, before being spun, and which are intended for the manufacturing of cloths and other mixed stuffs ; because the more allum there is in the liquor of the wool, the more it is harsh and difficult to spin, and it must spin thicker, and consequently the stuff is coarser. This observation is not of great consequence for spun wool which is intended for tapestry or for stuffs. I only mention it to shew that the quantity of ingredients may be sometimes varied without danger.

To avoid the harsh and brittleness of the wool, from the preparation that it receives from the earthly parts of these salts; step out of the old path, diminish the preparation as the shade requires, for every hundred weight of wool, use eight pounds of allum and three pounds of tartar; all yellows must undergo a preparation, and all colours connected with them. These rules for the preparation must be general for the yellow dye. To add when your dye is set and boiled ready to receive your wool or goods, take half a gill of the composition prepared for scarlet as will be described hereafter, to every hundred gallons of liquor; this may be added or diminished to the shade required ; it has a tendency to soften and enliven the body of the wool and make it pliable ; it is better than to load the bodies of the wool with these earthly and astringent salts, that leaves the wool harsh and brittle. There can be no objection to any astringent in any dye if properly applied ; it is an affinity on the wool to coat and make a body for the reception of the colouring substances ; the only objections, are the earth these preparatory salts contain.

I have given you the form of the preparation for the yellow, and its effects, I shall close with some observations on the colouring substances for yellow, as to the use and their connections with other colours, &c.

OF WELD.

WELD is a plant that may be cultivated among us, and is used grossly, either green or dry; but when in the blossom and not exposed to damage by the wet, and kept dry, it yields but little colour and is numbered the first in this dye ; to withstand all trials, it requires six pounds when dry, to every pound of wool or stuff, and boil well.

The yellow wood is used in chips, or in coarse shavings ; by this means it is more divided, and yields its dye the better, and a less quantity will do ; which way soever it is used, it is put into a bag, that it may not mix with the wool or stuffs. The same precaution is necessary for the savory and greenwood, when they are mixed with the weld to change its shade.

Greenwood commonly comes ground ; it gives a greenish yellow and is good in greens, olives and drabs; however, for browns sumac will produce the same effect, and when one cannot be obtained the other will answer.

The other ingredients are hitherto known which dye yellow, and I shall only observe here in regard to the good dye, that the root of the dock, the bark of the ash-tree, particularly that which is raised after the first sap, the leaves of almond, peach and pear-tree, the root and bark of the barbary shrub, saffron flowers, the herb peters-wort, and in particular the assmart, which gives a beautiful yellow if fermented before used in dying; its colour will be permanent. The woad in Europe, is prepared by a chymical process, and produces a large revenue; undoubtedly the assmart, which in the northern states is troublesome to farmers, might be a profit to them and our country, were it suitably prepared for a dye stuff; its extract is highly charged with acid and vegetable salts.

If our government should deem it worth their attention, to employ some able chymist to explore the qualities of our fossils, woods, barks, shrubs, plants, roots, weeds and minerals, perhaps the advantages our rising nation might derive, would soon indemnify us for the extra expense.

In short, all leaves, barks and roots, which by chewing shew some little astriction, give yellows of the good dye more or less fine, according to the time they are boiled, and in proportion to the tartar and allum used in the liquor: a proper quantity of allum brings these yellows to the beautiful yellow of the weld. If the tartar is in greater quantity, these yellows will border on the orange; and lastly, if these ingredients are too much boiled, let them be roots, barks, or leaves, the yellow obscures itself, and takes brown shades.

Although some dyers use turmeric in the good dye, which gives an orange yellow, this practice is to be condemned, for it is a colour that soon passes in the air, unless it be secured by sea-salt, which some dyers do, who take care to keep this imposition to themselves. Those who make use of it in common scarlets, to spare cochineal, and to give to their stuff a red bordering on the orange, are blameable, for the scarlets that have been dyed after this manner lose in a short time that bright orange; as I have already said, they brown considerably in the air. Yet these falsifications are obliged to be in some measure tolerated; for at this time that bright orange being in fashion, it would be impossible to give it to scarlet, without putting a larger dose of composition, whose acids would greatly hurt the cloth. The fustic wood is now preferred in scarlet.

OF TURMERIC.

THE turmeric is a root that is brought from the East Indies, which dyes a yellow; without it neither a good yellow, green or straw colour can be imprest on silks. Turmeric is a small root; if it be good, when broken it will be a dark yellow, have a strong flavour and be very bitter to the taste.

That which comes from Patna is most valued. The Indian dyers call it *haleh*; it is also called *concome* in the regulations of M. Colbert. It is reduced to a very fine powder, and used pretty near the same way as the grain of Avignon, but in much less quantity, on account of its yielding a great deal of dye. It is somewhat better than the other yellow ingredients that will be spoken of in the sequel, but, as it is dear, it is a sufficient reason for seldom or never using it in the lesser dye.

It is sometimes used in the great dye to gild the yellows made with weld, and to brighten and orange the scarlets; but this practice is to be condemned; for the air carries off all the colour of the turmeric in a short time, so that the gilded yellows return to their first state, and the scarlets brown considerably; when this happens to these sort of colours, it may be looked on as certain that they have been falsified with this ingredient, which is not lasting.

OF FUSTIC.

FUSTIC is much used in this country. The colour it naturally produces, is an orange yellow, and turns brown when long exposed to the air. It is employed in colouring saxon greens and olives; in short, it may be used in all colours where the ground requires a yellow; it is a clean wood, an astringent, and leaves the goods soft and pliable. There is not one among the ranks of the yellow materials that is so useful as the fustic for browns; as it changes it becomes darker and inclining to red, is useful in smokes, snuffs and cinnamon colours; it is good in black, and excellent in drabs. It is a close and hard wood, hard to split and full of splinters; the root and that part of the wood that is most knotty is the best, when split it should appear of a bright yellow; if it is rotten or otherwise injured it will not answer. Some condemn this wood because it is not good for the yellow, and will not tolerate it in the good dye; here are the reasons given by Mr. Haigh, dyer of Leeds:

"If a stuff dyed with fustic is dipped in the woad vat, a disagreeable olive ensues, which does not resist the air, but soon loses its colour." And that "fustic was made use of in Languedoc for making of lobster colours for foreign markets, as it greatly saves cochineal. For this purpose they mix weld, fustic, and cochineal, with a little cream of tartar, in the same liquor, and the stuff boiled in this liquor comes out of a lobster colour, and accordingly, to the quantity of these different ingredients, it becomes more or less red, tending to the orange. Although the method of mixing together ingredients of the good with those of the lesser dye ought to be condemned, yet in this case, and for this colour only, which is in considerable demand in the Mediterranean, it appears that the fustic may be tolerated ; for having attempted to make the same colour, with only the ingredients of the good dye, I did not get a more lasting colour.

"The change which the air produces in the lobster colour made with fustic is very sensible, but it is not so disagreeable as the changes incident to several other colours ; for all the shade goes off and weakens at once, so that it is rather diminution than a change of colour ; whereas the lobster colour made with the yellow wood becomes of a cherry colour"

It appears Mr. Haigh's remarks are groundless and without foundation, he condemns it for no other reason than because it does not answer all his purposes, yet gives it the preferance in the scarlet to the turmeric, and cannot well make the flame coloured scarlet without one or both of these substitutes, as those of the good dye give so little colour, that it will consume the red of the cochineal, &c.

Yellow oak bark produces a strong colour, green or dry, but it is better to have it roped and ground as for tanning ; it is also good in browns and blacks. Walnut or hickory bark may be used for the same colours ; it makes a brighter yellow than the oak, both are durable. In the use of some of these yellow subjects may be added a little blue vitriol to the dye, it will make it very brilliant and fine ; oil of vitriol may also be added, but it will not answer to make it general, only in cases of necessity, &c.

OF ROUCOU.

THE roucou or racourt is a kind of dry paste brought from America ; this ingredient gives an orange colour pretty near the same as the fustic, and the dye is not more lasting. However it is not by the proof allum that the quality of the roucou is to be judged, for this does not in the least alter its colour ; on the contrary, it becomes finer and brighter, but the air carries it off, and effaces it in a short time ; soap has the same effect, and it is by this it must be tried according to the instructions on these kind of proofs. The place of this ingredient is easily supplied in the good dye by weld and madder mixed together, but roucou is made use of in the lesser dye after the following manner.

Pearlash is dissolved in a copper with a sufficient quantity of water ; it is well boiled for one hour, that the ashes may be totally dissolved ; then as many pounds of roucou as there are of ashes, are added ; the liquor is well raked and suffered to boil for a quarter of an hour ; the wool or stuffs that are to be dyed are then dipped without any preparation, except dipping them in luke-warm water, that the colour may spread itself equally.

They are left in this liquor, working them continually until they are come to the desired shade, after which they are washed and dried.

The roucou is often mixed with other ingredients of the lesser dye, but I cannot give any instructions on these mixtures, as they depend on the shades you wish to make, and are in themselves attended with no difficulty.

I have boiled the stuff in allum and tartar before I dyed it with roucou, but though the colour was more lasting it was not sufficiently so to be deemed of the good dye. On the whole, the roucou is a very bad ingredient for dying of wool, and is not made much use of, for it is dear, and other ingredients, that are cheaper and hold better, are used in its stead.

Wool dyed with roucou, and afterwards dipt in the indigo or woad vat, take a reddish olive, which in a very short time becomes almost blue in the air, the colour given by the roucou disappearing.

Of the Grains of Avignon.

THE grains of Avignon are but little used in dying, they give a pretty good yellow, but not lasting, no more than the green, produced by dipping in the same liquor, a stuff that has a ground of blue. To work it, the stuff must be boiled in allum and tartar as for weld. Then a fresh liquor is made with these grains, and the stuff is dipt, and must lie in it longer or shorter, according to the shade that is wanted. There is no difficulty in working of it, so I need only observe that it ought never to be used but when all other ingredients for dying yellow are wanting; this must seldom happen, as they are neither scarce nor dear.

The yellows are easily obtained, more so than any other colour, but two simple processes are required; first, the preparation, then the dye and the colour required.

This is all that remains for me to say on the ingredients for yellow for the great or the lesser dye; the dye of the lesser dye is to be used for common and low-priced stuffs. It is not that I think it impossible to extract lasting colours from them, but then those colours will not strictly be the same which these ingredients yield naturally, or by the ordinary methods, as that gum and astriction which is wanting in them must be added, and then they are no more of the same quality ; consequently the rays of light will be differently reflected, and the colour will be different.

CHAP. III.

OF RED.

RED is one of the material or primitive colours, as has been before observed, and is dependent on the power of the acid always ; the alkali is sometimes admitted when the goods have received too much acid, and to change the red to a crimson. Crimson is considered by some as one of the reds, but I consider it as compounded, as you may see in the preceding work ; however it is so much connected with the reds, I shall class it with them ; violets, purples and all browns that the ground is red, are connected with the red, as will be shown in the sequel. Neutral substances are frequently introduced in the red dye, as verdigrease, blue vitriol, &c. these tend to sadden the goods, as the alkali, when they have received too much acid, and are bordering on the orange or yellow, and the red wants raising in the great dye : there are four principal reds, which are the basis of the rest, these are :

1. Scarlet of grain. 2. The scarlet, now in use, or flame-coloured scarlet, formerly called Dutch scarlet. 3. The crimson red. And, 4. The madder red.

There are also the bastard scarlet and the bastard crimson ; but as these are only mixtures of the principal reds, they ought not to be considered as particular colours.

The red, or *nacaret* of *bourre**, was formerly permitted in the great dye.

All these different reds have their particular shades from the deepest to the lightest, but they form separate classes, as the shades of the one never fall into those of the other.

The reds are worked in a different manner from the blues, the wool or stuffs not being immediately dipped in the dye, but previously receiving a preparation which gives them no colour, but prepares them to receive that of the colouring ingredient.

This is called the water of preparation ; it is commonly made with acids, such as sour waters, allum and tartar, aquafortis, aqua regalis, &c. These preparing ingredients are used in different quantities, according to the colour and shade required. Galls are also often used, and sometimes alkaline salts. This I shall explain in the course of this treatise, when I come to the method of working each of these colours.

It has been the opinion of some dyers, that the waters of America would not answer for a scarlet, and also that a vessel of silver or pure block tin was necessary to contain the scarlet dye; experience has taught us that these opinions are groundless, the waters of this country are as pure and soft as those of Europe ; a brass or copper caldron, if well cleansed will leave the colour as bright as any vessel whatever : brass is preferred, as it is easier kept clean, as may be seen in the preceding work of the different experiments in Europe. As scarlets are generally

*This colour is given with weld and goat's hair boiled in potash, and is a bright orange red.

dyed in the cloth, it is necessary they should be fulled and finished fit for the press, as soap will crimson it, and the hand, &c. would tarnish it in dressing.

Of Flame-Coloured Scarlet.

FLAME-coloured scarlet, that is, bright-coloured scarlet, known formerly under the name of Dutch scarlet, (the discovery of which Kunkel attributes to Kuster, a German chymist) is the finest and brightest colour of the dye. It is also the most costly, and one of the hardest to bring to perfection. It is not easy to determine the point of perfection, for independent of different tastes concerning the choice of colours, there are also general fancies, which make certain colours more in fashion at one time than another; when this happens, fashionable colours become perfect ones. Formerly scarlets were chosen full, deep, and of a degree of brightness which the sight easily bore. At this time they must be on the orange, full of fire, and of a brightness which dazzles the eye. I shall not decide which of these two fashions deserve the preference, but shall give the method of making them both, and all the shades which hold a medium between these extremes.

Cochineal, which yields this beautiful colour, and is also called mestick, or tescalle, is an insect that is gathered in great quantities in Mexico. The natives and Spaniards, who have but small establishments there, cultivate them, that is, carefully gather them from the plant on which they feed before the rainy season. They kill and dry those designed for sale and preserve the rest to multiply when the bad season is over. The insect feeds and breeds upon a kind of prickly opuntia, which they call topal.

It may be preserved in a dry place for ages without spoiling.

The cochineal sylvestre, or campessiane, is also brought from Vera-Cruz. The Indians of Old and New Mexico gather this kind in the woods; it feeds, grows, and generates there on the wild uncultivated opuntias; it is there exposed in the rainy season to all the humidity of the air, and dies naturally. This cochineal is always smaller than the fine or cultivated; the colour is more holding and better, but has not the same brightness, neither is it profitable to use it, since it requires four parts, and sometimes more, to do what may be done with one of fine.

Sometimes they have damaged cochineal at Cadiz; this is fine cochineal that has been wet with salt water, occasioned by some shipwreck or leakage. These accidents considerably diminish the price, the sea-salt saddening the dye. This kind serves only to make purples, and even those are not the best. However a person in 1735, found the secret to turn this to almost as much advantage for scarlet as the finest cochineal. The discovery of this secret is easy, but let him that possesses it enjoy it, I shall not deprive him of the advantage he might have in it.

Almost every dyer has a particular receipt for dying scarlet, and each is fully persuaded that his own is preferable to all others; yet the success depends on the choice of the cochineal, of the water used in the dye, and on the manner of preparing the solution of tin, which the dyers call composition of scarlet.

As it is this composition which gives the bright flame colour to the cochineal dye, and which without this acid liquor would naturally be of a crimson colour, I shall describe the preparation that succeeded best with me, and then point out the different processes as practised in Europe and their success, and opinions in the manner of using the preparations and applying the colouring substances. We are furnished in the good or great dye, with four colouring substances for red, the kermes, the cochineal, gum lacque and madder, there is a number in the false dye, as red-wood or brazil, nicaragua, &c.

Receipt 128th. For Scarlet, as practised in America.

WHEN your cloth or goods are prepared for dying, to every fourteen pounds weight take twelve ounces of cochineal, ten ounces of cream of tartar, two pounds of double aquafortis, two ounces of salts of sal ammoniac, two ounces of sal nitre or salt petre, six pounds of wheat bran, two ounces of turmeric and six ounces of granulated tin.

Composition for Scarlet.

TAKE twelve pounds of double aquafortis, to which add gradually twelve pounds of clear clean water, put in a large glass vessel; then add three-quarters of a pound of salts of sal ammoniac made fine, put it in gradually, then take three quarters of a pound of sal nitre or salt petre, pulverized and added slowly, shake them all together till the salts are all dissolved, then add two pounds and a quarter of granulated tin, dropping it in by little and little, as it dissolves it will cause a great fermentation, and you must not be in too great haste in adding the tin; when the tin is all in and the ebullition ceases, then stop it tight with a glass or wax stopper, put it where it may not be disturbed or shaken up, for the sediment will injure the dye, let stand for use; it must be prepared twenty four hours before using: if you keep it stopped close you may keep this composition good several months; this is the composition for scarlet.

To prepare or granulate the Tin.

TAKE of the purest block tin or grain tin, that is a metal by itself; it comes in various sizes, from half an ounce to one pound in weight, it has a bright appearance. Take the tin and melt it over a hot fire, then hold it two feet above a pail of clean cold water, and pour it gradually into the water, then take it out and dry it for use.

The cloths and composition all prepared, then clean the copper clean as described in the preceding work, have all the dying utensils new and clean, or that have not been used in any other dye; then fill with fair water and clean, and your goods clean and wet with clean water; to fourteen pound weight of cloth, take six pounds of wheat bran, put it in a clean linen bag tied close, boil it two hours; then take up the bag, let it drain, then take twelve ounces cochineal and ten ounces of cream of tartar, have it well pulverized together in a glass or marble mortar and glass pestel, sifted through gauze; when thus prepared, add one-third of this compound of cochineal, &c. to the boiling liquor, run your goods two hours boiling, turning lively, then take up and air; this is called the scarlet boiling; then shift the liquor from your copper, fill with clean water, and heat boiling hot; then add half of the remaining compound of cochineal, and two pounds and four ounces of the composition, carefully turned off that you get none of the sediment, for that will injure the dye, boil well, run your cloth one hour with the dye boiling, tend lively, air, and add the remainder of the cochineal, &c. and as much more of the composition as before, and two ounces of turmeric made fine, boil well and run as before. If your dye stuff and composition are good, your cloth and utensils clean, you will have as good a scarlet as was ever made in Europe. I can vouch for this form being used with success in the United States, and was equal to any scarlet I ever saw.

Receipt 129*th*. *Of Scarlet of Grain.*

THIS colour is called scarlet of grain, because it is made with the kermes, which was long thought to be the grain of the tree on which it is found. It was formerly called French scarlet, imagining it to be first found out in France, and is now known by the name of Venetian scarlet, being much in use there, and more made than in any other place. The fashion passed from thence into France and other countries. It has indeed less lustre, and is browner than the scarlet now in fashion; but it has the advantage of keeping its brightness longer, and does not spot by mud or acid liquors.

The kermes is a gall insect, which is bred, lives, and multiplies upon the *ilex acculeato cocci glandiscra, C. B. P.* Some comes from Narbonne, but greater quantities from Alicant and Valentia, and the peasants of Languedoc yearly bring it to Montpelier and Norbonne. The merchants who buy them to send abroad, spread them on cloths, and sprinkle them with vinegar, in order to kill the little insects that are within, which yield a red powder, which is separated from the shell after drying, and is then passed through a sieve; this is done particularly in Spain.

They then makeit up in bales, and in the middle of each a quantity of this powder is inclosed in a leather bag, in proportion to the whole bale. Thus each dyer has his due proportion of this powder. These bales are generally sent to Marseilles, from whence they are exported to the Levant, Algiers, and Tunis, where it is greatly made use of in dying.

The red draperies of the figures in the ancient tapestry of Brussels, and other manufactories of Flanders, are dyed with this ingredient; and some that have been wrought upwards of two hundred years, have scarcely lost any thing of the brightness of the colour. I shall now proceed to give the method of making this scarlet of grain, which is now seldom used but for wools designed for tapestry.

Preparation of the wool for Scarlet of Grain.

Twenty pounds of wool and half a bushel of bran are put into a copper, with a sufficient quantity of water, and suffered to boil half an hour, stirring it every now and then; it is then taken out to drain.

It is necessary to observe, that whenever spun wool is to be dyed, a stick is passed through each hank (which commonly weighs one pound) and they remain on the stick during the course of the work to prevent their entangling. This stick also enables the dyer to return the hanks with more ease, by plunging each part successively in the liquor, by which they take an equal dye; by raising the hank with a stick, and drawing it half way out of the copper, seizing the other end of the hank with the other hand, it is plunged towards the bottom. If the wool be too hot, this may be done with two sticks, and the oftener this is repeated, the more even will be the dye; the ends of the sticks are then placed on two poles to drain. These poles are fixed in the wall above the copper.

Liquor for the Kermes.

WHILE this prepared wool is draining, the copper is emptied, and fresh water put in, to which is added about a fifth of sour water, four pounds of Roman allum grossly powdered, and two pounds of red tartar. The whole is brought to boil, and that instant the hanks are dipped in (on the sticks) which are to remain in for two hours, stirring them continually one after the other after the method already laid down.

I must in this place observe, that the liquor in which the allum is put, when on the point of boiling sometimes rises so suddenly that it comes over the copper, if not prevented by adding cold water. If, when it is rising, the spun wool is instantly put in, it stops it, and produces the same effects as cold water.

The liquor does not rise so suddenly when there is a large quantity of tartar, as in the process; but when the allum is used alone, sometimes above half the liquor comes over the copper when it begins to boil, if not prevented by the method described.

When the wool has boiled two hours in this liquor, it is taken out, left to drain, gently squeezed, and put into a linen bag in a cool place for five or six days, and sometimes longer; this is called leaving the wool in preparation. This is to make it penetrate the better, and helps to augment the action of the salts, for as a part of the liquor always flies off, it is evident that the remaining, being fuller of saline particles, becomes more active, provided there remained a sufficient quantity of humidity; for the salts being crystalized and dry, would have no more action.

I have dwelled much longer on this preparing liquor, and the method of making it, than I shall in the sequel, as there are a great number of colours for which it is prepared pretty near in the same proportion, so that when this happens, I shall slightly describe it, mentioning only the changes that are to be made in the quantity of allum, tartar, sour water, or other ingredients.

After the spun wool has been covered five or six days, it is fitted to receive the dye. A fresh liquor is then prepared according to the quantity of wool to be dyed, and when it begins to be lukewarm, take 12 ounces of powdered kermes for each pound of wool to be dyed, if a full and well-coloured scarlet is wanted. If the kermes was old and flat, a pound of it would be required to each pound of wool. When the liquor begins to boil, the yarn (still moist, which it will be if it has been well wrapped in the bag, and kept in a cool place) is put in. If it had been boiled a long time before, and grown dry, it must be lightly passed through lukewarm water, and well squeezed before it is dyed.

Previous to its being dipped in the copper with the kermes, a handful of wool is cast in, which is let to boil for a minute : this takes up a kind of black scum, which the kermes cast up, by which the wool that is afterwards dipped acquires a finer colour. This handful of wool being taken out, the prepared is to be put in. The hanks are passed on sticks as in the preparation, continually stirring, and airing them one after the other. It must boil after this manner an hour at least, then taken out and placed on the poles to drain, afterwards wrung and washed.

The dye still remaining in the liquor, may serve

to dip a little fresh parcel of prepared wool; it will take some colour in proportion to the goodness and quality of the kermes put into the copper.

When different shades are wanted, a less quantity of kermes is used, so that for twenty pounds of prepared wool seven or eight are sufficient.

The quantity of wool that is to have the lightest shade is first to be dipped, and to remain no longer in than the time sufficient to turn it and make it take the dye equally. Then the next deepest shade intended is dipped, and left to remain some time longer; after this manner the work is continued to the last, which is left as long as requisite to acquire the necessary shade.

The reason of working the lightest shades first, is, that if the yarn is left too long in, no damage is done, as that hank may serve for a deeper shade; whereas, if they begin by a deeper, there would be no remedy if a failure happened in some of the lighter shades. The same caution is to be taken in all colours whose shades are to be different.

There are seldom more shades than one from the colour now spoken of; but as the working part is the same for all colours, what has been said on this subject will serve for the rest.

The yarn thus dyed, before bringing it to the river, may be passed through lukewarm water, in which a small quantity of soap has been perfectly dissolved: this gives a brightness to the colour, but at the same time saddens it a little, that is, gives it a little cast of the crimson. As I shall often make use of the terms *rouzing* and *saddening*, especially in the acids, it is necessary to explain their meaning.

Saddening, is giving a crimson or violet cast to red; soap and alkaline salts, such as lie of ashes, potash, lime, sadden reds; thus they serve to bring them to the shade required when too bright, and that they are too much rouzed.

Rouzing, is doing quite the reverse; it is giving a fire to the red, by making it border on the yellow or orange. This is performed on wool by the means of acids, as red or white tartar, cream of tartar, vinegar, lemon juice, and aquafortis. These acids are added more or less, according to the depth of the orange colour required. For example, if the scarlet of grain was wanted to be more bright, and approach somewhat nearer to common scarlet, a little of the scarlet composition, which has been spoken of, must be poured into the liquor after the kermes is put in, and the brown colour of that liquor would immediately be brightened by the acid, and become of a brighter red; the wool dipped in would be more liable to be spotted by mud and acid liquors: the reason will appear in the next chapter.

I have made various experiments on this colour, in order to make it fitter and brighter than what it generally is, but I never could extract a red that was to be compared to that of cochineal.

Of all the liquors which I made for the preparation of the wool, that which was made with the preparations just mentioned succeeded best. By changing the natural dye of the kermes, by different kinds of ingredients of metallic solutions, &c. various colours are made, which I shall immediately speak of.

I shall say but little about dying stuffs with this red, as the proportion cannot be prescribed for each yard of stuff, on account of their breadth and thickness, or the quantity of wool entering their composition; practice alone will teach the necessary quantity for each sort of stuff; however, not to work in the dark, or to try experiments at random, the surest way will be to weigh the stuffs, and to diminish about one-fourth part of the colouring ingredients laid down for spun wool, as stuffs take up less colour inwardly, their texture being more compact, prevents its penetration, whereas yarn or wool in the fleece receives it equally within and without.

The allum and tartar for the liquor of preparation for the stuffs must be diminished in the same proportion, and they are not to remain in the preparing liquor as long as the wool. It may be dyed the next day after boiling.

If wool in the fleece is dyed with the red of the kermes, either to incorporate it with cloths of a mixed colour, or to make full cloths, it will have a much finer effect than if the wool had been dyed in the red of madder. I shall mention this in describing the compound colours in which the kermes is used, or ought at least be used in preference to madder, which does not give so fine a red, but, being cheaper, is commonly substituted for it.

Half grain scarlet, or bastard scarlet, is that which is made of equal parts of kermes and madder. This mixture affords a very holding colour, not bright, but inclining to a blood red. It is prepared and worked in the same manner as that made of kermes alone. This dye is much cheaper and the dyers commonly make it less perfect by diminishing the kermes and augmenting the madder.

By the proofs that have been made of scarlet of grain or kermes, whether by exposing it to the sun, or by different proofs, it is certain there is not a more holding or a better colour; yet the kermes is no where in use but at Venice. The mode of this colour has been entirely out since the making of flame-coloured scarlets. This scarlet of grain is now called a colour of bullock's blood: nevertheless, it has great advantages over the other, for it neither blackens nor spots, and grease may be taken out without prejudice to its colour; but it is out of fashion and that is sufficient. This has entirely put a stop to the consumption of kermes in France. Scarce a dyer knows it, and when Monsieur Colbert wanted a certain quantity for the experiments above related, he was obliged to send for it to Languedoc, the merchants of Paris keeping only a sufficiency for medicinal purposes.

When a dyer is obliged to dye a piece of cloth, known yet under the name of scarlet of grain, as he has neither the knowledge of the kermes, nor the custom of using it, he makes it of a cochineal, as I shall relate in the following receipt; it comes dearer, and is less holding than that made of the kermes. The same is done in regard to spun wool designed for tapestries, and as this shade is pretty difficult to hit with cochineal, they commonly mix brazil wood, which hitherto has been a false ingredient, permitted only in the lesser dye. For this reason all these kind of reds fade in a very short time, and though they are much brighter than required, coming out of the hands of the workman, they lose all their brightness before the expiration of a year: they whiten or become exceeding grey; it is therefore to be wished that the use of kermes was again established.

It is also certain, that if some dyer set about using it, there are several colours that might be extracted from it with more ease and less expence than the common method; for these colours would be better and more holding, and he would thereby acquire a greater reputation. I have made above fifty experiments with the kermes, from which some use in practice may arise; I shall only relate such as have produced the most singular colours.

By mixing the kermes with cream of tartar, without allum, and as much of the composition as would be used for the making of scarlet with cochineal, you have in one liquor an exceeding bright cinnamon, for nothing but the acid entering in the mixture, the red parts of the kermes become so minute that they almost escape the sight. But if this cinnamon colour be passed through a liquor of Roman allum, part of this red appears again; whether it be by the addition of the allum that drives out a part of the acid of the composition, or the earth of the allum precipitated by the astriction of the kermes, which has the effect of galls, I know not; but this red thus restored is not fine.

With cream of tartar (the composition for scarlet) and allum, in greater quantity than tartar, the kermes gives a lilac colour, which varies according as the proportion of ingredients are changed.

If in the place of allum and tartar, ready prepared tartar of vitriol is substituted, which is a very hard salt resulting from the mixture of the vitriolic acid and a fixed alkali, such as the oil of tartar potash, &c. and if, I say, after boiling the kermes in a solution of a small quantity of this salt, the stuff be dipped in and boiled one hour, it acquires a tolerable handsome agath grey, and in which very little red is seen, for the acid of the composition having too much divided the red of the kermes, and the tartar of vitriol, not containing the earth of the allum, it could not re-unite these red atoms, dispersed by precipitation. These agath greys are of the good dye, for, as I have observed in the chapter treating of indigo, the tartar of vitriol is a hard salt, which is not calcined by the sun, and is indissoluble in rain water.

Glauber salts mixed with the kermes entirely destroy its red, and give an earthy grey that does not stand the proof, for this salt neither resists cold water nor the rays of the sun, which reduce it into powder. Vitriol or green copperas, and blue vitriol separated substituted for allum, but joined to the crystal of tartar, equally destroy or veil the red of the kermes, which in these two experiments produce the same effect as if galls or sumac had been made use of; for it precipitates the iron of the green vitriol, and dyes the cloth of a grey brown, and the copper of the blue vitriol dyes it of an olive.

Instead of blue vitriol, I used a solution of copper* in aquafortis, which also produced an olive colour; a convincing proof that the kermes has the precipitating quality of the galls, since it precipitates the copper of the vitriol as a decoction of gall-nut would.

There is great probability that what renders the red of the kermes as holding as that of madder, is from the insects feeding on an astringent shrub, which, notwithstanding the changes made by the digestion of the juices of the plant, still retains the astringent quality of the vegetable, and consequently the virtue, and so gives a greater spring to the pores of the wool to contract themselves quicker and with greater strength, when it comes out of the boiling water, and is exposed to the cold air; for I have observed that all barks, roots, wood, fruits, and other matters that have some astriction, yield colours of the good dye.

* Verdigrease.

Receipt* 130*th*. *Flame coloured scarlet, as practised at Leeds and in France.

Composition for Scarlet —Take eight ounces of spirit of nitre, (which is always purer than the common aquafortis mostly used by the dyers) and* be certain that it contains no vitriolic acid ; weaken this nitrous acid by putting it into eight ounces of filtered river water : dissolve in it, little by little, half an ounce of very white salt ammoniac, to make it an aqua regia, because spirits of nitre alone will not dissolve block-tin. Lastly, add two drachms of salt petre ; this might be omitted, but I observed that it was of use in making the dye smooth and equal. In this aqua regia thus weakened, dissolve one ounce of the best block-tin, which is first granulated or made small while melted by casting it from a height into a vessel of cold water. These small grains of tin are put into the dissolvent one by one, letting the first dissolve before putting in others ; this prevents the loss of the red vapours, which would rise in great abundance, and be lost if the dissolution of the metal was made too hastily ; it is necessary to preserve these vapours, and, as Kunkel observed, they greatly contribute towards the

* Dissolve in a small quantity of spirit of nitre as much silver as it will take ; put a few drops of this into some of the spirit of nitre that is to be proved ; if this spirit remains transparent, it is pure ; but if a white cloud be perceived, which will afterwards form a sediment, it is a sign that there is a commixture of vitriol or spirit of salt. In order therefore to render the spirit of nitre absolutely pure, drop the solution of silver gradually into it, so long as it shall produce the least turbidness, time being given for the spirit to become clear betwixt each addition. The spirit of nitre being then poured off from the sediment, will be perfectly pure ; and if this sediment, which is the silver precipitated, be evaporated to dryness, and then infused in a crucible with a small quantity of any fixed alkaline salt, it will be reduced to its proper metalline state.

brightness of the colour, either because these vapours are acids that evaporate and are lost, or contain a sulphur peculiar to salt petre, which gives a brightness to the colour. This method is indeed much longer than that used by the dyers, who immediately pour the aqua-fortis upon the tin reduced to small pieces, and wait till a strong fermentation ensues, and a great quantity evaporates before they weaken it with common water. When the tin is thus dissolved, this scarlet composition is made, and the liquor is of the beautiful colour of dissolved gold, without any dirt or black sediment, as I used very pure tin without allay, and such as runs from the first melting of the furnaces of Cornwall. This solution of tin is very transparent when newly made, and becomes milky and opaque during the great heat of summer; the greatest part of the dyers are of opinion, that it is then changed and good for nothing; yet mine, notwithstanding this defect, made as bright scarlet as if it had remained clear; besides, in cold weather what I made recovered its first transparency. It must be kept in a glass bottle with a stopper, to prevent the evaporation of the volatile parts.

As the dyers do not attend to this, their composition often becomes useless at the end of twelve or fifteen days. I have laid down the best method, and, if they seek perfection, they will abandon their old practice, which is imperfect.

The dyers in France first put into a stone vessel, with a large opening, two pounds of salt ammoniac, two ounces of refined saltpetre, and two pounds of tin reduced to grains by water, or, which is still preferable, the filings of tin; for when it has been melted and granulated, there is always a small portion converted into a calx which does not dissolve. They weigh four pounds of water in a separate vessel, of which they pour about two ounces upon the mixture in the stone vessel ; they then add a pound and a half of common aquafortis, which to it produces a violent fermentation. When the ebullition ceases, they put in the same quantity of aquafortis, and an instant after they add one pound more. They then put in the remainder of the four pounds of water they had set aside ; the vessel is then close covered, and the composition let to stand till the next day.

The salt petre and salt ammoniac are sometimes dissolved in the aquafortis before the tin is put in ; they practise both methods indiscriminately, though it is certain that this last method is best. Others mix the water and aquafortis together, and pour this mixture on the tin and salt ammoniac. In short, every dyer follows his own method.

Water for the Preparation of Scarlet.

The day after preparing the composition, the water for the preparation of scarlet is made, which differs from that made in the preceding receipt.

Clear the water well. For each pound of spun wool, put twenty quarts of very clear river water (hard spring water will not do) into a small copper. When the water, is a little more than lukewarm, two ounces of the cream of tartar finely powdered, and one drachm and a half of powdered and sifted cochineal is added. The fire is then made a little stronger, and when the liquor is ready to boil two ounces of the composition are put in. This acid instantly changes the colour of the liquor, which, from a crimson, becomes of the colour of blood.

As soon as this liquor begins to boil, the

wool is dipped in, which must have been previously wetted in warm water and wrung. The wool is continually worked in this liquor, and left to boil an hour and a half; it is then taken out, slightly wrung, and washed in fresh water. The wool coming out of the liquor is of a lively flesh colour, or even some shades deeper, according to the goodness of the cochineal, and the strength of the composition. The colour of the liquor is then entirely passed into the wool, remaining almost as clear as common water.

This is called the water of preparation for scarlet, and the first preparation it goes through before it is dyed; a preparation absolutely necessary, without which the dye of the cochineal would not be so good.

Reddening.

To finish it, a fresh liquor is prepared with clear water, the goodness of the water being of the greatest importance towards the perfection of the scarlet. An ounce and a half of starch is put in*, and when the liquor is a little more than lukewarm, six drachms and a half of cochineal finely powdered and sifted is thrown in. A little before the liquor boils, two ounces of the composition is poured in, and the liquor changes its colour as in the former. It must boil, and then the wool, is put into the copper, and continually stirred as in the former. It is likewise boiled an hour and a half; it is then taken out, wrung, and washed. The scarlet is then in its perfection.

One ounce of cochineal is sufficient for a pound of wool, provided it be worked with at-

* Starch softens it.

tention, and after the manner laid down, and that no dye remains in the liquor. For coarse cloth less would do, or half as much for worsted. However, if it was required to be deeper of cochineal, a drachm or two might be added, but not more, for it would then lose its lustre and brightness.

Though I have mentioned the quantity of the composition, both in the water of the preparation and the dye, yet this proportion is not to be taken as a fixed rule.

The aquafortis used by the dyers, is seldom of an equal strength; if, therefore, it be always mixed with an equal quantity of water, the composition would not produce the same effect: but there is a method of ascertaining the degree of acidity of aquafortis. For example, to use that only, two ounces of which would dissolve one ounce of silver. This would produce a composition that would be always equal, but the quality of the cochineal would then produce new varieties, and the trifling difference that this commonly causes in the shade of scarlet is of no great signification, as more or less may be used to bring it precisely to the colour desired. If the composition be weak, and the aforesaid quantity not put in, the scarlet will be a deeper and fuller in colour. On the contrary, if a little more is added, it will be more on the orange, and have what is called more fire; to rectify which, add a little of the composition, stirring it well in the copper, having first taken out the wool; for if it was to touch any part before it was thoroughly mixed, it would blot it. If, on the contrary, the scarlet has too much fire, that is, too much on the orange, or too much rouzed, it must be passed through clear warm water; when finished, this saddens it a little, that is, diminishes its bright orange; if there still remained too much, a little Roman allum must be mixed with the hot water.

For spun wool that is to have all the various shades of scarlet, about half the cochineal, and half the composition for full scarlet is sufficient. The cream of tartar must also be diminished proportionably in the water of preparation. The wool must be divided into as many hanks or skains as there are to be shades, and when the liquor is prepared, the skains that are to be lightest are first to be dipped, and to remain in but a very short space of time; then those that are to be a little deeper, which must remain in somewhat longer, and thus proceeding to the deepest; the wool is then to be washed, and the liquor prepared to finish them. In this liquor, each of these shades are to be boiled one after the other, beginning always with the lightest, and if they are perceived not to be of the proper shade they must be passed again through the liquor. The eye of a dyer, will readily judge of the shades, and a little practice will bring this to perfection.

The dyers are divided in opinion of what metal the boiler should be made. In Languedoc they use those made of the finest block-tin, and several dyers, in Paris follow the same method. Yet that great dyer, M. de Julienne, whose scarlets are in great repute, uses brass. The same is used in the great manufactory at St. Dennis. M. de Julienne, to keep the stuffs from touching the boiler, makes use of large rope nets with close meshes, At St. Dennis, instead of a rope net, they have large baskets, made of willow stripped of the bark, and not too close worked.

As so much had been said concerning the metal of the boiler, I tried the experiment. I took two ells of white sedan cloth, which I dyed in two separate boilers of equal size; one was of brass, fitted with a rope net, the other of block tin. The cochineal, the composition, and other ingredients, were weighed with the utmost accuracy and boiled precisely the same time. In short, I took all possible care that the process should be the same in both, that if any difference arose it might only be attributed to the different metals of the boiler. After the first liquor, the two pieces of cloth were absolutely alike only that which had been boiled in the tin vessel appeared a little more streaked and uneven, which, in all likelihood, proceeded from these two ells of cloth being less scoured at the mill than the two others; the two pieces were finished each in the separate boilers, and both turned out very fine; but that which had been made in the tin boiler had a little more fire than the other, and the last was a little more saddened. It would have been an easy matter to have brought them both to the same shade, but that was not my intention.

From this experiment, I conclude, that when a brass boiler is used, it requires a little more of the composition than the tin one; but this addition of the composition makes the cloth feel rough; to avoid this defect, the dyers who use brass vessels put in a little turmeric, a drug of the dye, but which gives to scarlet that shade which is now in fashion; I mean that flame-colour, which the eye is scarce able to bear.

This adulteration is easily discovered by cutting a piece of the cloth; if there is no turmeric, the web will be of a fine white, but yellow if there is. When the web is dyed the same as the surface, it is said that colour is webbed, and the contrary, when the middle of the weaving remains white. The lawful scarlet is never dyed in the web: the adulterated, where the turmeric or fustic has been made use of, is more liable to change its colour in the air than the other. But as the brightest scarlets are now in fashion, and must have a yellow cast, it is better to tolerate the use of turmeric, than to use too great a quantity of the composition to bring the scarlet to this shade; for in this last case, the cloth would be damaged by it, would be sooner spotted by dirt from the quality of the acid, and would be more easily torn, because acids stiffen the fibres of the wool, and render them brittle.

I must also take notice, that if a copper vessel is used it cannot be kept too clean. I have failed several times with my patterns of scarlet, by not having the copper scoured.

I cannot help condemning the common practice of some dyers, even the most eminent, who prepare their liquor over night, and keep it hot till next morning, when they dip in their stuffs; this they do, not to lose time, but it is certain that the liquor corrodes the copper in that space, and by introducing particles of copper in the cloth, prejudices the beauty of the scarlet. They may say they only put in their composition just at the time when the cloth is ready to be dipt in the copper; but the cream of tartar, or the white tartar, which they put in over night, is an acid salt sufficient to corrode the copper of the vessel, and form a verdigrease, although it dilutes itself as it forms, still has not a less effect.

It would therefore be better to make use of tin boilers, a boiler of this metal must contribute to the beauty of scarlet; but these boilers of a sufficient size cost much, and may be melted by the negligence of the workmen, and there is a difficulty in casting them of so great a size without sand flaws, which must be filled. Now if these sand-holes are filled with solder, there must of necessity be places in the boiler that contain lead; this lead in time being corroded by the acid of the composition, will tarnish the scarlet. But if such a boiler could be cast without any sand-holes, it is certain such a one would be preferable to all others, as it contracts no rust, and if the acid of the liquor detaches some parts, they cannot be hurtful.

Having laid down the manner of dying spun wool in scarlet, and its various shades, which are so necessary for tapestry and other work, it is proper to give an idea of the dying of several pieces of stuff at one time. I shall relate this operation as it is practised in Languedoc. I made the trial on some ells of stuff, which succeeded very well, but this scarlet was not so fine as the flame coloured.

There are two reasons why the wool is not dyed before it is spun (for fine colours) first in the course of the manufacturing, that is, either in the spinning, carding, or weaving, it would be almost impossible in a large workshop, where there are many workmen, but that some particles of white wool, or some other colour would mix, which would spoil that of the stuff by blotting it ever so little; for that reason, the reds, the blues, the yellows, the greens, and all other colours that are to be perfectly uniform, are never dyed before they are manufactured.

The second reason, which is peculiar to scarlet, or rather to cochineal, is, that it will not stand the milling, and as the greatest part of high stuffs must be milled after they are taken from the loom, the cochineal would lose part of its colour, or at least would be greatly saddened by the soap, which produces this effect by the alkaline salt which destroys the brightness given to the red by the acid. These are the reasons that the cloths and stuffs are not dyed in scarlet, light red, crimson, violet, purple, and other light colours, but after being entirely milled and dressed.

To dye, for example, five pieces of cloth at one time of five quarters breadth, and containing fifteen or sixteen ells each, the following proportions are to be observed. Put into a stone or glazed earthen pot twelve pounds of aquafortis, and twenty pounds of water, to which add a pound and a half of tin, made in grains by running it in water, or filed. The dissolution is made quicker or slower, according to the greater or lesser acidity of the aquafortis. The whole is left to rest twelve hours at least, during which time a kind of black mud settles at the bottom of the vessel; what swims over this sediment is poured off by inclination; this liquor is clear and yellow, and is the composition which is to be kept by itself.

This process differs from the first in the quantity of water mixt with the aquafortis, and in the small quantity of tin, little of which must remain in the liquor, since aquafortis alone cannot dissolve it, but only corrodes it, and reduces it to a calx, as there is neither salt petre, not salt ammoniac which would form an aqua regia. However, the effect of this composition differs from the first only to the eyes accustomed to judge of that colour.

This composition made without salt ammoniac, and which has been of long use amongst a great number of manufacturers at Carcassone, who certainly imagined that its effect was owing to the sulphur of the tin, can only keep thirty-six hours in winter without spoiling, and twenty-four hours in summer; at the expiration of which it grows muddy, and a cloud precipitates to the bottom of the vessel, which changes to a white sediment. This is the small quantity of tin, which was suspended in the acid, but an acid not prepared for that metal; the composition which ought to be yellow becomes at this time as clear as water, and if used in that state would not succeed; it would have the same effect as that which would become milky.

The late M. Baron pretended to have been the first discoverer at Carcassone of the necessity of adding salt ammoniac to hinder the tin from precipitating. If so, there was no one in that town that knew that tin cannot be really dissolved but by aqua regia.

Having prepared the composition as I have described it after M. de Fondriers, about sixty cubical feet of water are put into a large copper for the five pieces of cloth before mentioned, and when the water grows warm, a bag with bran is put in, sometimes also sour waters are used; the one and the other serve to correct the water, that is, to absorb the earthy and alkaline matters which may be in it, and which, as I have already said, saddens the dye of the cochineal, for the effect of the water ought to be well known, and experience will teach whether such expedients should be used, or whether the water, being very pure and denulated of salts and earthy particles, can be used without such helps.

Be that as it will, as soon as the water begins to be little more than lukewarm, ten pounds of powdered cream of tartar is flung in, that is, two pounds for each piece of cloth. The liquor is then raked strongly, and when it grows a little hotter, half a pound of cochineal is cast in which is well mixt with sticks; immediately after twenty-seven pounds of the composition very clear is poured in, which is also well stirred, and as soon as the liquor begins to boil, the cloths are put in, which are made to boil strongly for two hours, stirring them continually by the help of the wynch: they are then taken out upon the scray, and well handled three or four times from end to end, by passing the lists between the hands to air and cool them. They are afterwards washed.

After the cloth has been washed, the copper is emptied and a fresh liquor prepared, to which if necessary, a bag with bran or some sour water is added; but if the water is of a good quality, these are to be omitted; when the liquor is ready to boil, eight pounds and a quarter of powdered and sifted cochineal is put in, which is to be mixed as equally as possible throughout the liquor, and having left off stirring, it is to be observed when the cochineal rises on the surface of the water, and forms a crust of the colour of the lees of the wine; the instant this crust opens of itself in several places, eighteen or twenty pounds of the composition is to be added. A vessel with cold water must be at hand to cast on the liquor in case it should rise, as it sometimes does, after the composition is put in.

As soon as the composition is in the copper, and equally distributed throughout the whole, the cloth is cast in, and the wynch strongly turned two or three times, that all the pieces may equally take the dye of the cochineal. Afterwards it is turned slowly to let the water boil, which it must do very fast for one hour, always turning the wynch, and sinking the cloth in the liquor with sticks, when by boiling it rises too much on the surface. The cloth is then taken out, and the lists passed between the hands to air and cool it; it is then washed, after which it is to be dried and dressed.

In each piece of the Languedoc scarlet cloth there is used, as has been shewn, one pound and three-quarters of cochineal in the dye and preparation; this quantity is sufficient to give the cloth a very beautiful colour. If more cochineal was added, and a deeper orange-colour required, the quantity of the composition must be augmented.

When a great quantity of stuffs are to be dyed in scarlet, a considerable profit arises by doing them together, for the same liquor serves for the second dip which was used for the first. For example: when the five first pieces are finished, there always remains in the liquor a certain quantity of cochineal, which in seven pounds may amount to twelve ounces; so that if this liquor be used to dye other stuffs, the cloths dipped in it will have the same shade of rose colour as if they had been dyed in a fresh liquor with twelve ounces of cochineal; yet this quantity may vary pretty much, according to the quality or choice of the cochineal, or according to the fineness it has been reduced to when powdered. But whatever colour may remain in the liquor, it deserves some attention on account of the high price of this drug. The same liquor is then made use of for other five pieces, and less cochineal and composition are put in proportion to what may be judged to remain; fire and time are also saved by this, and rose-colour and flesh-colour may also be produced from it; but if the dyers have no leisure to make these different liquors in twenty-four hours, the colour of the liquor corrupts, grows turbid, and loses the rose-colour entirely. To prevent this corruption some put in Roman allum, but the scarlets which are prepared after that manner are ill saddened.

When cloths of different qualities, or any other stuffs are to be dyed, the surest method is to weigh them, and for each hundred weight of cloth add about six pounds of crystal or cream of tartar, eighteen pounds of composition in the water of preparation, as much for the reddening, and six pounds and a quarter of cochineal. Thus in proportion for one pound of stuff use one ounce of cream of tartar, six ounces of composition, and one ounce of cochineal; some eminent dyers at Paris put two-thirds of the composition and a fourth of the cochineal in the water of preparation, and the other third of the composition with three-fourths of the cochineal in reddening.

It is not customary to put cream of tartar in the reddening, yet I am certain, by experience, that it does not hurt, provided the quantity does not exceed half the weight of the cochineal, and it appeared to me to make a more lasting colour. Some dyers have made scarlet with three dippings; namely, a first and second water for preparation, and then the reddening; but still the same quantity of drugs is always used.

I observed, in the foregoing receipt, that the little use made of kermes for the brown or Venetian scarlets, obliges most dyers to make them with cochineal; for this purpose a water of preparation is made as usual; and for the reddening, eight pounds of allum are added for each hundred weight of stuff; this allum is dissolved by itself in a kettle, with a sufficient quantity of water, then poured into the liquor before the cochineal is put in. The remainder is performed exactly as in the common scarlet; this is the Venetian scarlet, but it has not near the same solidity as if made with the kermes.

There are no alkaline salts which do not sadden scarlet; of this number are the salt of tartar, potash, pearlash calcined, and nitre fixed by fire; therefore allum is more generally used; and if these alkaline salts be boiled with the stuffs, they would considerably damage them, for they dissolve all animal substances. If the allum be calcined, it is still the more secure.

The redder the scarlet is, the more it has been saddened; from thence it appears that these colours lose in the liquor that browns them a part of their ground ; however one cannot brown in the good dye but with salts. The late M. Baron observes, in a memoir he gave sometime ago to the Royal Academy of Sciences, that all the salts he had made use of for browning, making the colour smooth, and preserving its brightness and deepness, he had succeeded best with salt of urine, but, as he observes, it is too troublesome to make this salt in any quantity.

I said, in the preceding receipt and the chapter on yellow, that the choice of the water for scarlet and other bright colours was very material, and as the greatest part of common water saddens it, for they mostly contain a chalky, calcareous earth, and sometimes a sulphureous or vitriolic acid ; these are commonly called hard waters, that is, they will not dissolve soap or boil vegetables well. By finding a method of absorbing or precipitating these hurtful matters, all waters may be equally good for this kind of dye : thus, if alkaline matters are to be removed, a little sour water produces this effect ; for if five or six buckets of these sour waters are mixed with sixty or seventy of the hard water before it comes to boil, these alkaline earths rise in a scum, which is easily taken off the liquor.

All that I have hitherto said in this chapter is for the instruction of dyers ; I shall now make an attempt to satisfy the philosopher how these different effects are produced.

Cochineal, infused or boiled by itself in pure water, gives a crimson colour bordering on the purple ; this is its natural colour ; put it into a glass, and drop on it spirits of nitre ; this colour will become yellow, and if you still add more, you will scarcely perceive that there was originally anyred in the liquor; thus the acid destroys the red by dissolving it and dividing its parts so minutely that they escape the sight. If in this experiment a vitriolic, instead of a nitrous acid be used, the first changes of the colour will be purple, then purpled lilac, after that a light lilac, then flesh-colour, and lastly colourless. This blueish substance, which mixes with the red to form a purple, may proceed from that small portion of iron, from which oil of vitriol is rarely exempt. In the liquor of preparation for scarlet, no other salt but cream of tartar is used, no allum is added as in the common preparing water for other colours, because it would sadden the dye by its vitriolic acid; yet a calx or lime is required, which, with the red parts of the cochineal, may form a kind of lake, like that the painters use, which may set in the pores of the wool by the help of the crystal of tartar.

This white calx is found in the solution of very pure tin, and if the experiment of the dye is made in any small glazed earthen vessel, immediately on the cochineal's communicating its tincture to the water, and then adding the composition drop by drop, each drop may be perceived with a glass or lens, to form a small circle, in which a brisk fermentation is carried on; the calx of the tin will be seen to separate, and instantaneously to take the bright dye, which the cloth will receive in the sequel of the operation.

A further proof that this white calx of tin is necessary in this operation, is that if cochineal was used with aquafortis, or spirits of nitre alone a very ugly crimson would be obtained; if a solution of any other metal was made use of in spirits of nitre, as of iron or mercury, from the first would be had a deep cinder-grey, and from the second, a chesnut colour with green streaks, without being able to trace in the one or other any remains of the red of the cochineal. Therefore, by what I have laid down, it may be reasonable to suppose, that the white calx of the tin, having been dyed by the colouring parts of the cochineal, rouzed by the acid of the dissolvent of this metal, has formed this kind of earthy lake whose atoms have introduced themselves into the pores of the wool, which were opened by the boiling water, that they are plaistered by the crystal of tartar, and these pores, suddenly contracting by the immediate cold the cloth was exposed to by airing, that these colouring particles are found sufficiently set in to be of the good dye, and that the air will take off the primitive brightness, in proportion to the various matters with which it is impregnated. In the country, for example, and particularly if the situation be high, a scarlet cloth preserves its brightness much longer than in great cities, where the urinous and alkaline vapours are more abundant. For the same reason, the country mud with, which in roads is generally but an earth diluted by rain water does not stain scarlet as the mud of towns where there are urinous matters, and often a greatdeal of dissolved iron, as in the streets of great cities, for it is well known that any alkaline matter destroys the effect which an acid has produced on any colour whatsoever. And for the like reason, if a piece of scarlet is boiled in a lie of potash, this colour becomes purple, and by a continuation of boiling it is entirely taken out ; thus from this fixed alkali, and the crystal of tartar, a soluble tartar is made, which the water dissolves and easily detaches from the pores of the wool : all the mastic of the colouring parts is then destroyed, and they enter into the lies of the salts.

Receipt 131*st*. *Scarlet of Gum-Lacque.*

THE red part of the gum-lacque may be

also used for the dying of scarlet, and if this scarlet has not all the brightness of that made of fine cochineal alone, it has the advantage of being more lasting.

The gum-lacque, which is in branches or small sticks and full of animal parts, is the fittest for dying. It must be red within, and its external parts of a blackish brown; it appears by a particular examination made of it by M. Geoffroy some years since, that it is a sort of hive, somewhat like that of bees, wasps, &c.

Some dyers make use of it powdered and tied in a linen bag; but this is a bad method, for there always passes through the cloth some resinous portion of the gum, which melts in the boiling water of the copper, and sticks to the cloth, where it becomes so adherent when cold, that it must be scraped off with a knife.

Others reduce it to powder, boil it in water, and after it has given all its colour, let it cool, and the resinous parts fall to the bottom. The water is poured out, and evaporated by the air, where it often becomes stinking, and when it has acquired the consistence of thick honey, it is put into vessels for use. Under this form it is pretty difficult justly to determine the quantity that is used; this induced me to seek the means of obtaining this tincture separated from its resinous gum, without being obliged to evaporate so great a quantity of water to have it dry, and to reduce it to powder.

I tried it with weak lime water, with a decoction of the heart of agaric, with a decoction of comfrey-root, recommended in an ancient book of physic; in all these the water leaves a part of the dye, and it still passes too full of colour, and it ought to be evaporated to get all the dye; this evaporation I wanted to avoid, therefore I made use of mucilaginous or slimy roots, which of themselves gave no colour, but whose mucilage might retain the colouring parts, so that they might remain with it on the filter.

The great comfrey-root has, as yet, the best answered my intention: I use it dry and in a gross powder, putting half a drachm to each quart of water, which is boiled a quarter of an hour, passing it through a hair sieve. It immediately extracts from it a beautiful crimson tincture; put the vessel to digest in a moderate heat for twelve hours, shaking it seven or eight times to mix it with the gum that remains at the bottom, then pour off the water this is loaded with colour in a vessel sufficiently large, that three-fourths may remain empty, and fill it with cold water: then pour a very small quantity of strong solution of Roman allum on the tincture; the mucilaginous or slimy dye precipitates itself, and if the water which appears on the top appears still coloured, add some drops of the solution of allum to finish the precipitation, and this repeat till the water becomes as clear as common water.

When the crimson mucilage or slime is all sunk to the bottom of the vessel, draw off the clear water, and filter the remainder; after which, dry it in the sun.

If the first mucilaginous water has not extracted all the colour of the gum-lacque, (which is known by the remaining being of a weak straw colour) repeat the operation until you separate all the dye the gum-lacque can furnish; and as it is reduced to powder when dry, the quantity to be used in the dye is more exactly ascertained than by evaporating it to the consistence of an extract.

Good gum-lacque, picked from its sticks, yields, dried and powdered, but little more dye than one-fifth of its weight. Thus at the price it bears at present, there is not so great an advantage as many may imagine in using it in the place of cochineal; but to make the scarlet colour more lasting than it commonly is, it may be used in the first liquor or preparation, and cochineal for reddening.

If scarlet is made of gum-lacque, extracted according to the method here taught, and reduced to powder, a caution is to be taken in dissolving it, which is useless when cochineal is used; that is, if it was put into the liquor ready to boil, the dyer would lose three-quarters of an hour, before it would be dissolved entirely; therefore for despatch, put the dose of this dry tincture into a large earthen vessel, or into one of tin, pour warm water on it, and when it is well moistened, add the necessary dose of the composition for scarlet, stirring the mixture well with a glass pestel. This powder, which was of a dirty deep purple, as it dissolves takes fire-coloured red extremely bright; pour the dissolution into the liquor, in which was previously put the crystal of tartar, and as soon as this liquor begins to boil, dip the cloth in, keeping it continually turning. The remaining part of the operation is the same as that of scarlet with cochineal: the extract of gum-lacque, prepared according to my method, yields about one-ninth more of dye than cochineal, at least than that which I made use of for this comparison.

If instead of the crystal of tartar and the composition of some fixed alkaline salt or lime water is substituted, the bright red of the gum-lacque is changed into the colour of lees of wine, so that this dye does not sadden so easily as that of cochineal.

If instead of these alternatives, salt ammoniac

is used by itself, cinnamon or clear chesnut colours are obtained, and that according as there is more or less of this salt. I have made twenty other experiments on this drug, which I shall not relate here, because they produced none but common colours, and which may be easier had from ingredients of a lower price. My experiments were with a view of improving the red of the lacque, and the method I have here laid down to extract its colouring parts answers extremely well; the more ingredients that are discovered for scarlet, the less will be the cost; for, although these experiments made on cochineal, lacque, and other drugs may appear useless to some dyers, they will not be so to others who study to improve this art.*

Receipt 132*d*. *Of the Red of Madder*.

THE root of madder is the only part of this plant which is used in dying. This plant may be cultivated in the United States of America to great advantage; it is three years after the first root is set in the ground before it comes to maturity, or the ground filled with roots fit for digging or breaking up; if it remains in the ground longer than three seasons, there will be a quantity of useless roots; they may be placed four feet apart, in the first setting in the ground, and hoed the first year to keep it clear from weeds; if the ground has a deep soil it will be filled with small roots to the depth of three feet; it yields abundantly; the time of drying, which is in autumn, in the month of October, or the last of September, spade up the earth, take the

* The colouring parts of the gum-lacque may be extracted by common river water, by making it a little more than lukewarm, and inclosing the powdered lacque in a coarse woollen bag.

roots from it, assort them carefully, and wash them clean in cold water and lay them to dry for manufacturing. The small bright and young roots that have no bark nor pith, are for the good or grape-madder.

Of all the reds this is the most lasting, when it is put on a cloth or stuff that is throughly scoured, then prepared with the salts with which it is to be boiled two or three hours, without which, this red, so tenacious after the preparation of the subject, would scarcely resist more the proofs of the reds than any other ingredients of the false dye. This is a proof that the pores of the fibres of the wool ought not only to be well scoured from the yolk or unctuous transpiration of the animal, which may have remained, notwithstanding the scouring of the wool after the common manner with water and urine; but it is also necessary, that these same pores be plaistered inwardly with some of those salts which are called hard, because they do not calcine in the air, and cannot be dissolved by rain water, or by the moisture of the air in rainy weather. Such is, as has been said before, the white crude tartar, the red and the crystal of tartar, of which, according to common custom, about a fourth is put into the preparing liquor, with two-thirds or three-fourths of allum.

The best madder roots come generally from Zealand, where this plant is cultivated in the islands of Tergoés, Zerzée, Sommerdyke, and Thoolen. That from the first of these islands is esteemed the best; the soil is clay, fat, and somewhat salt. The lands that are deemed the best for the cultivation of this plant are new lands, that only served for pasture, which are always fresher and moister than others. The Zealanders are beholden to the refugees of

Flanders for the cultivation and great commerce of this root.

It is known in trade and dying under the names of grape-madder, bunch-madder, &c. It is however the same root; all the difference in regard to its quality is, that the one kind contains pith and root, and the other has the small fibres from its pricipal root adhering to it.

Both are prepared by the same work, which I shall not relate the particulars of here, as it would only serve to lengthen this treatise to no purpose.

They choose the finest roots for the first sort, drying them with care, grinding them and separating the rind at the mill, and preserving the middle of the root ground in hogsheads, where it remains for two or three years; for after this time, it is better for dying than it would have been coming from the mill; for if madder was not kept close after this manner, the air would spoil it, and the colour would be less bright. It is at first yellow, but it reddens and grows brown by age; the best is of a saffron colour, in hard lumps, of a strong smell, and yet not disagreeable. It is also cultivated about Lisle in Flanders, and several other places of the kingdom, where it was found to grow spontaneously.

The madders which are made use of in the Levant and in India, for the dying of cottons, are somewhat different from the kinds used in Europe, it is named *chat* on the coast of Coromandel. This plant thus called, grows abundantly in the woods on the coast of Malabar, and this chat is the wild sort. The cultiv ted comes from Vasur and Tuccorin, and the most esteemed of all is the chat of Persia, named *dumas*.

They also gather on the coast of Coromandel the root of another plant called *ray de chaye*, or root of colour, and which was thought to be a kind of *rubia tinctorum*, but is the root of a kind of *gallium flore albo*, as it appeared by observations sent from India in 1748. It has a long slender root, which dyes cotton of a tolerable handsome red, when it has received all the preparations previous to the dye.

At Kurder, in the neighbourhood of Smyrna, and in the countries of Akissar and of Yordas, they cultivate another kind of madder, which is called in the country *chioc-boya ekme hazala*. This of all the madders is the best for the red dye, by the proofs that have been made of it, and far more esteemed in the Levant than the finest Zealand madder the Dutch bring there. This madder so much valued is called by the modern Greeks *lizari*, and by the Arabs *foiioy*.*

There is another kind of madder in Canada called *tyssa-voyana*. It is a very small root, which produces pretty near the same effect as the European madder.

The water of preparation for madder red is pretty near the same as for kermes, that is composed of allum and tartar. The dyers do not agree as to the proportions; but the best appears to be four ounces of allum and one of red tartar to each pound of spun wool, and about one-twelfth part of sour water, and let the wool boil in it for two hours. If it is spun wool, leave it for seven or eight days, that it may be well moistened by the dissolution of these salts; and if it is cloth, finish it the fourth day.

To dye wool with madder, prepare a fresh liquor, and when the water is come to a heat to

* These kinds of madders give brighter reds than the best grape-madder of Zealand, for they are dyed in the air and not in a stove. The madder of Languedoc, even that of Poitou, succeeds as well as that of lizari, when it is dryed witheut fire.

bear the hand, put in half a pound of the finest grape-madder for each pound of wool; let it be well raked and mixed in the copper before the wool goes in, keep the wool in an hour, during which time it must not boil.* Shades from madder are obtained after the manner laid down for other colours, but these shades are little used, except in a mixture of several colours.

When several pieces of cloth are to be dyed at once in madder red, the operation is the same, as you may see in the 29th receipt in the preceding for red with madder, only augmenting the ingredients in proportion; and let it be remarked that in small operations the quantity of ingredients must be somewhat greater than in great, not only in madder red, but in all other colours.

These reds are never so beautiful as those of the kermes, and much less so than those of the lacque or cochineal, but they cost less, and are made use of for common stuffs whose low prices would not allow a dearer dye. Most of the reds for the army are of madder, saddened with archil or brazil, (though these drugs be of the safe dye) to make them finer, and more on the velvet, which perfection could not be procured to them even with cochineal, without considerably augmenting the price.

I have already said that madder put on stuffs not being prepared to receive it by the allum and tartar-water, did in fact give its red colour, but that which it dyed was blotted and not lasting, it is therefore the salts that secure the dye; this is common to all other colours red or yellow, which cannot be made without a preparing liquor. Now the question is, whether these act by taking off the remains of the oily and fat

* If madder is boiled, its red becomes obscure, and of a brick colour.

transpiration of the sheep, or whether that of the two salts, particularly that which even cannot be carried by luke-warm water, remains to catch, seize and cement the colouring atom, opened or dilated by the heat of water to receive it, and contracted by the cold to retain it.

To determine which, use any alkaline salts, such as potash, the clarified lays of oak-ashes, or any other pure lixivial salt instead of allum and tartar, put in a due proportion so as not to dissolve the wool, and afterwards dip the stuff in madder liquor. This stuff will come out coloured, but will not last, even boiling water will carry off three-fourths of the colour. Now it cannot be said that a fixed alkaline salt is unfit to extract from the pores of the wool the yolk or fat of the sheep, since lixivial salts are used with success in several cases, to take the grease out of stuffs of what kind soever they be, which water alone could not take off. It is also well known, that with fats foreign to the stuff, and an alkaline salt, a kind of soap is formed which water easily carries off.

Again, take a piece of stuff dyed in madder red, according to the usual method, boil it some time in a solution of fixed alkaline salts, a small quantity will also destroy the colour, for the fixed alkali, attacking the small atoms of the crystal of tartar, or crude tartar, which lines the pores of the wool, forms a soluble tartar, which water dissolves very easily, and consequently the pores being opened in the hot water of the experiment, the colouring atom came out with the saline atom that sheathed it.

This stuff being washed in water, the remaining red colour is diluted, and a colour half brown and half dirty remains. If instead of an alkaline salt, soap is substituted, (which is an alkaline salt, mitigated by oil) and another piece of cloth dyed also in madder, be boiled for a few minutes, the red will become finer, because the alkali which is in the soap being sheathed with oil, it could not attack the vegetable acid, and the boiling only carried off the colouring parts ill stuck together, and their numbers diminishing, what remains must appear deeper or clearer.

I must also add, for further proof of the actual existence of salts in the pores of a stuff prepared with allum and tartar, before dying it with madder, that more or less tartar gives an infinite variety of shades with this root only; for if the quantity of allum be diminished, and that of the tartar augmented, a cinnamon will be had, and even if nothing but tartar alone be put into the liquor, the red is lost, and a deep cinnamon or brown root colour is obtained, though of a very good dye; for the crude tartar, which is an acid salt, has so much dissolved the part which should have produced the red colour, that there only remained a very small quantity, with the ligneous fibres of the root, which, like all other common roots, does then yield but a brown colour, more or less deep according to the quantity used. I have already proved that the acid which brightens the red, dissolves them if too much is used, and divides them into particles so extremely minute, that they are not perceptible.

If in the place of tartar, any salt which is easily dissolved be put with the allum in the liquor, to prepare the stuff for the madder dye, such as salt petre, the greater part of the madder red becomes useless, it disappears, or does not stick on, and nothing is got but a very bright cinnamon, which will not sufficiently stand the proof, because the two salts used in the preparing liquor are not of the hardness of the tartar.

Volatile urinous alkalis which are obtained from certain plants, such as the perilla, the archil of the Canaries, and other mosses or lichens, destroy also the madder red, but at the same time communicate another to it, for on experiment, madder prepared after the manner of archil with fermented urine and quick lime, produced only nut colours, but which nevertheless are lasting; because there entered into the liquor only the little portion of urinous volatile that moistened the madder which the boiling was sufficient to evaporate, and besides, the cloth was sufficiently furnished with the salts of the liquor made as usual, to retain the colouring parts of the dye.

When a pure red, that for cochineal an example, is laid upon a cloth first dyed in blue, and afterwards prepared with the liquor of tartar, and allum to receive and retain this red, a purple or violet is produced according to the quantity of blue or red. The red of madder has not this effect, for it is not a pure red like that of the cochineal, and as I said above, it is altered by the brown ligneous fibres of its root, and makes on the blue a chesnut colour, more or less deep according to the preceding intensity of the blue first laid on. If this chesnut colour is wanted to have purple cast, a little cochineal must be added.

In order to avoid this brown of the root, the dyers who make the best reds of madder take great heed to use the liquor of madder a little more than luke-warm; the madder tarnishes considerably by the heat of the water, extracting the particles which dye brown, and unite themselves with the red.

This inconveniency might be remedied, if at the time that the madder root is fresh a means could be found to separate from the rest of this root the red circle which is underneath its brown pelicle, and which surrounds the middle pith ; but this work would augment its price, and even then it would not afford so good a red as cochineal. However, it might be attempted to dye cottons red, whose price might bear the expenses of this preparation.

Madder being of all ingredients the cheapest of any that dye red and of the good dye, it is mixt with others to diminish the price. It is with madder and kermes that the bastard scarlets of grain are dyed, otherwise called half-grain scarlets, and with madder and cochineal the half-common scarlets, and the half-crimsons are made.

To make the half-grain scarlet, the water of preparation, and all the rest of the operation is to be performed after the same manner as scarlet made of the grain of kermes, or the common Venetian, only the second liquor is composed of half kermes and half grape-madder.

For the half-scarlet and flame-colour, the composition and preparation is as usual, nothing but pure cochineal being put in, but in the reddening, half cochineal and half madder is used : here also the sylvestre may be made use of, for after having made the preparation with cochineal, for reddening, use half a pound of cochineal, a pound and a half of sylvestre, and one pound of madder instead of cochineal alone.

That the wool and stuffs may be dyed as equally as possible, it is necessary that the two kinds of cochineal be well rubbed or sifted, as also the madder, with which they must be well incorporated before they are put into the liquor. This must be observed in all colours where several ingredients are mixt together. This half-scarlet is finished like the common scarlet, and it may be saddened after the same manner, either with boiling water or allum.

The half-crimson is made like the common crimson, only using half madder, and half cochineal, the cochineal sylvestre may be used here also, observing only to retrench half of the common cochineal, and to replace it with three times as much of the sylvestre. If a greater quantity of the sylvestre was used, and more of the other taken off, the colour would not be so fine. Various shades may be produced by augmenting or lessening the madder or cochineal.

Receipt 133d. For Crimson.

CRIMSON, as I have already observed, is the natural colour of the cochineal, or rather, that which it gives to wool boiled with allum and tartar, which is the usual water of preparation for almost all colours. This is the method which is commonly practised for spun wool; it is almost the same for cloths, as will be seen hereafter.

For each pound of wool, two ounces and a half of allum, and an ounce and a half of white tartar, are put into the copper. When the whole boils, the wool is put in, well stirred, and left to boil for two hours; it is afterwards taken out slightly wrung, put into a bag, and left thus with its water, as for the scarlet in grain, and for all other colours.

For the dye a fresh liquor is made, in which three-fourths of an ounce of cochineal is added for each pound of wool. When the liquor is little more than luke-warm, the cochineal is put in, and when it begins to boil, the wool is cast in, which is to be well stirred with sticks; it is to remain thus for an hour; when taken out, wrung and washed.

If degrees of shades are required, (whose names are merely arbitrary) proceed, as has been already related for the scarlet, using but half the cochineal at first, and beginning with the lightest.

The beauty of crimson consists in its bordering as much as possible on the grisdelin, a colour between a grey and a violet. I made several trials to bring crimson to a higher perfection than most dyers have hitherto done, and indeed I succeeded so as to make it as fine as the false crimson, which is always brighter than the fine.

This is the principle on which I worked. As all alkalis sadden cochineal, I tried soap, barilla, potash, pearlash; all these salts brought the crimson to the shade I wanted, but at the same time, they tarnished and diminished its brightness. I then bethought myself to make use of volatile alkalis, and I found that the volatile spirit of salt ammoniac produced a very good effect; but this spirit instantly evaporated, and a pretty considerable quantity was used in the liquor, which greatly augmented the price of the dye.

I then had recourse to another expedient which succeeded better, the expense of which is trifling. This was to make the volatile alkali of the salt ammoniac enter into the liquor, at the very instant that it comes out of its basis; and to effect this, after my crimson was made after the usual manner, I passed through a fresh liquor, in which I had dissolved a little of the salt ammoniac. As soon as the liquor was a little more than lukewarm, I flung in as much potash as I had before of salt ammoniac, and my wool immediately took a very brilliant colour.

This method even spares the cochineal; for this new liquor makes it rise, and then less may be used than in the common process; but the greatest part of dyers, even the most eminent, sadden their crimsons with archil, a drug of the false dye.

Very beautiful crimsons are also made by boiling the wool as for the common scarlet, and then boiling it in a second liquor, with two ounces of allum and one ounce of tartar, for each pound of wool, leaving it one hour in the liquor. A fresh liquor is then prepared, in which six drachms of cochineal is put for every pound of wool. After it has remained an hour in this liquor, it is taken out, and passed immediately through a liquor of barilla and salt ammoniac. By this method, gradations of very beautiful crimson shades are made by diminishing the quantity of the cochineal. It is to be observed, that in this process there are but six drachms of cochineal to dye each pound of wool, because in the first liquor a drachm and a half of cochineal is used for each pound. It is also necessary to remark, that, to sadden these crimsons, the liquor of the alkaline salt and salt ammoniac be not made too hot, because the separation of the volatile spirit of this last salt would be too quick, and the crystal of tartar of the first liquor would lose its proper effect by being changed, as I have already said into a soluble tartar.

The same operation may be done by using one part of the cochineal sylvestre instead of the fine cochineal, and the colour is not less beautiful, for commonly four parts of sylvestre have not more effect in dying than one part of fine cochineal. The sylvestre may also be used in dying scarle, but with great precaution; it should only be used in bastard scarlets and half-crimsons. I shall speak of this when I treat of these colours in particular.

When a scarlet is spotted or spoiled in the

operation by some unforeseen accident, or even when the dye has failed, the common remedy is to make it a crimson, and for that purpose, it is dipt in a liquor where about two pounds of allum are added for each hundred weight of wool. It is immediately plunged in this liquor, and left there until it has acquired the shade of the crimson desired.

Receipt 134th. For Languedoc Crimson.

I shall now shew the method they follow in Languedoc to make a very beautiful sort of crimson, or the cloths exported to the Levant, but which is not so much saddened as that which I have just spoken of, and which resembles much more the Venetian scarlet. For five pieces of cloth, the pieces are 25 yards when milled of broad cloth one and a half yards wide the liquor is prepared as usual, putting bran if necessary. When it is more than lukewarm, ten pounds of sea-salt are put, instead of crystal of tartar, and when it is ready to boil, twenty-seven pounds of the scarlet composition, made after the manner of carcassine already described, are poured in, and without adding cochineal the cloth is passed through this liquor for two hours, keeping it always turning with the wynch, and continually boiling. It is afterwards taken out, aired and washed; then a fresh liquor is made, with eight pounds and three-quarters of cochineal powdered and sifted, and when it is ready to boil, twenty-one pounds of composition are added; the cloth is boiled for three quarters of an hour with the common precautions, after which it is taken out, aired and washed: It is of a very fine crimson, but very little saddened; if it is required to be more saddened, a greater quantity of allum is put into the first liquor of preparation, and in the second less of the composition, the sea-salt is also added to this second liquor; a little practice in this method will soon teach the dyer to make all the shades that can properly be derived from crimson.

Whenever cochineal has been used, there is found at the bottom of the reddening liquor a quantity of very brown sediment, which is flung away with the liquor as useless. I examined it and found, that the liquor for the reddening of scarlet contained a precipitated calx of tin: I united this metal with a great deal of trouble; the remaining parts of this sediment are the dross of the white tartar, or of the cream of tartar, united with the gross parts of the bodies of the cochineal, which is, as has already been said, a small insect. I washed these little animal parts in cold water, and, by shaking this water, I collected, with a small sieve, what the agitation caused to raise on the surface.

After this manner I separated these light parts from the earthy and metallic; I dried them separately, then levigated them with equal weight of fresh crystal of tartar; I boiled a portion with a little allum, and put in a pattern of white cloth, which boiled for three quarters of an hour, at the end of which it was dyed of a very beautiful crimson.

This experiment having convinced me, that by powdering and sifting the cochineal as is commonly practised, all the profit that might be extracted from this dear drug is not obtained, I thought proper to communicate this discovery to the dyers, that they might avail themselves of it by the method following.

Take one ounce of cochineal powdered and sifted as usual; mix with it a quarter of its weight of very white cream of tartar very crystaline and very airy; put the whole on a hard levigating stone, and levigate this mixture till it is reduced to an impalpable powder; make use of this cochineal thus prepared in the liquor, and in the reddening, subtracting from the cream of tartar, which is to be used in the liquor, the small quantity before used with the cochineal. What is put to the reddening, although mixed with a fourth of the same salt, does not prejudice its colour, it even appeared to me that it was more solid. Those that will follow this method will find that there is about a fourth more profit to be obtained by it.

Receipt 135th. The Natural Crimson in Grain.

In proportion for every pound of cloth or other things, take two ounces of tartar pure, and two ounces of allum; boil them with the goods an hour and a half; then rince the goods very well from the boiling. The kettle must be filled again with clear water and a few handfuls of bran put in, in order to take out the filth of the water, as well as to soften it. Scum the scurf off when it begins to boil, and put in an ounce of well powdered grain, with one drachm of red arsenic and one spoonful of burnt wine lees; this gives a pretty lustre; then wash and rince it well, and you have most beautiful colour.

Receipt 136th. Scarlet; of the dying of flock or goat's hair.

THERE are two preparations very different one from the other in the dying of flock: the first is with madder, and belongs to the great and good dye; the second is to dissolve it and make use of it; this belongs to the lesser dye. The dying with flock was formerly permitted in the good dye, but was rather on account of its being extracted from madder, than by any experiment that had been made concerning its durability. I tried it with great attention, and found it beyond any doubt that there is no colour that resists the air less. It is certainly for this reason that it was restrained to the lesser dye in the new regulation of France in 1737. Yet, as by the same regulation, it is not permitted to the dyers of the lesser dye to use madder, nor even to keep it in their houses; it has been enacted, that only the dyers of the great dye should be suffered to madder flock, and those of the lesser dye to dissolve and use it.

To madder the flock or goat's hair, four pounds of either of them is cut and well separated, that the dye may penetrate the better. It is boiled two hours in a sufficient quantity of sour water; then it is drained for an hour, and put into a middling copper, half filled with water, with four pounds of roach allum, two pounds of red tartar, and one pound of madder. The whole is boiled for six hours, putting in hot water as the liquor wastes; it is left all night and next day in this liquor; the third day it is taken out and drained in a basket. Some dyers let it remain eight days, but it often happens that by this delay in a copper vessel it is tarnished by the liquors corroding a part of the copper; a middling copper is then filled to the two-thirds with half sour water, and half common water and when the liquor is ready to boil, eight pounds of madder, well cut and crushed between the hands, is added. When the madder is well mixed in the liquor, four pounds of flock or hair is put in and boiled for six hours; it is then well washed, and the next day it is maddered a second time after the same manner, only putting in four pounds of madder instead of eight, which were before used. After this second maddering, it is well washed and dried ; it is then almost black and fit for use.

It appears by this operation, that four pounds of flock or hair is loaded with thirteen pounds of the dye of madder, yet there still remains some dye in the liquor, which is then called an old maddering, and which is preserved for use on certain occasions, as in tobacco, cinnamon colour, and several others.

When the flock is thus maddered by the dyer of the great dye he sells it to dyers of the lesser, who have then the liberty to dissolve and use it ; this is the common method, which has many difficulties, and is known but to few dyers. Madder is hereby made fine.

About half an hour after seven in the morning six pails full of clear water are put into a middling copper, and when the water is lukewarm, five pounds of pearlash are put in : the whole is boiled till eleven, and the liquor is then considerably diminished, so as to be held in a lesser copper, into which it is emptied, observing first to let the dregs of the pearlash subside, that none but the clear may be used.

A pail full of this liquor is afterwards put into the middling copper, having first scoured it well, and a little fire made under it ; the four pounds of maddered flock are scattered in by degrees, and at the same time a little of the lukewarm and saline liquor of the small copper is added to keep down the boiling, which rises from time to time to the top of the copper, in which the operation is performing.

When all the flock and the liquor of the little copper are put into the middling one, a pail full of clear water is put on the dregs of the pearlash remaining in the little copper. This water serves to fill the middling one as the liquor in it evaporates. All this flock melts, or is dissolved by the action of the pearlash, and after the first half hour, not the least hair is to be perceived. The liquor is then of a very deep red. The whole is then boiled without any addition, till three in the afternoon, that the whole dissolution of the flocks may be the more exactly performed. Then a stick is placed upon the copper, and upon this stick is placed a pail of fermented urine, in which pail a small hole has been previously made towards its lower part, and a little straw put into it, that the urine may very slowly run into the copper; whilst it is running, the liquor is made to boil strongly, and this urine makes good what may be lost by evaporation. This operation continues five hours, during which time three pails full of urine are discharged into the copper, being made to run faster when the boil is stronger, than when moderate. It is here to be observed, that, on account of the small quantity of flock in the experiment which I lay down here, five pounds only of pearlash are ordered; for when thirty pounds of flock are dissolved at one time which is the common custom of the French dyers, they put twelve ounces of pearlash to each pound of flock.

During the whole time of this operation, a strong volatile smell of urine is emitted, and there swims on the surface of the liquor a brown scum, but much more so after the addition of the urine. The liquor is known to be sufficiently done when this rises no more, and that the boil rises but gently, that is what happened to the operation now related, at eight in the evening. The fire is then raked out, the copper covered, and thus left to the next day. Patterns had been taken at different times of the colours of the liquor from three to eight in the evening, by dipping in small pieces of paper: the first were very brown, and they became continually lighter, and they united themselves more and more, in proportion as the volatile part of the urine acted on the colouring parts of the liquor.

Nothing now remained but to dye the wool in the liquor thus prepared, and which is called melting of flock; this is the easiest work belonging to the dyer. A quarter of an hour before the dying is begun, a little piece of very clean roach allum is put in, and the copper is well raked to melt it. As this liquor which was in the middling copper had been covered the whole night, and the fire had not been put out, the liquor was still so hot as not to suffer the hand. The clearest was taken out and brought into a small copper, with a sufficient quantity of luke warm water, some wool dyed yellow with weld was dipped in it; it immediately became of a fine orange, bordering on the flame colour, that is of the colour called *nacaret*, and known to the dyers by the name of *nacaret of flock*, because it is commonly made with melted flock.

Twenty hanks of white wool were dipped one after the other in the same liquor, beginning by those that were to have the deepest ground, and leaving them longer or shorter in the liquor according to the shade required. An assortment was made after this manner from the nacaret, or bright orange red, to the cherry colour. It ought to be observed, that in proportion as the liquor was consumed, fresh was taken from the middle sized copper, great care being taken not to stir the sediment at the bottom; a little fire was also kept under the small copper, to keep the liquor always in the same degree of heat. The wool is thus dipped until the whole liquor is used, and all the colour drawn out. But the lighter colours could not be dyed in it; for when the colour of the liquor is once weakened, as it ought to be for these colours, it is generally loaded with filth, which would take off the brightness required in these shades.

The following is the method of making shades lighter than the cherry colour. A copper is filled with clear water, and five or six hanks of wool dyed of the deepest dye from the flock, that is, from the shade that immediately follows the nacaret, are put in. As soon as the water boils, it takes out all the colour the wool had, and it is in this fresh liquor that the other wool that is to be dyed is dipped, from the cherry colour to the palest flesh colour, observing always to begin by the deepest shades.

Most of the dyers who do not know how to melt the flock, or who will not give themselves that trouble, buy some pounds of this scarlet of flock, which they use after this manner, to make all the lighter shades, which, as has been said, is done with much ease. This operation shows what little dependance can be put on the solidity of a colour that passes so quickly in boiling water. And in fact, it is one of the worst colours there is in dying, and on that account the new regulation has taken it from the great dye, and permits in the lesser for the reason above mentioned.

Thus a very bad colour may be had from an ingredient which, of all those that are used in dying, is perhaps the best and the most durable; yet when this hair, dyed with all the necessary precautions to insure the colour as much as possible, comes to be dissolved or melted in a liquor of pearlash, its colour, by acquiring a new lustre, loses all its solidity, and can only be ranked in the number of the falsest dyes.

It may appear that the little solidity of this colour proceeded from the wool having no preparation, and retaining no salt before its being dipped in the dissolved flock; but I found that this was not the cause; for I dipped in this liquor wool boiled as usual, and other wool differently prepared, without finding that the colour of the latter had acquired any more solidity; the lustre was less, that is, it came out more saddened than the wool that had been dyed in it without any preparation.

Though I have said that wool receives no preparation before its being dyed in a dissolution of flock, it is nevertheless necessary to sulphur those that are to make clear shades, for that gives them a great brightness and lustre, as the dissolved flock is applied on a ground a great deal whiter than it would be without the vapour of the sulphur, which cleanses it of all its filth. The same thing is done for the light blues, and for some other colours; but this operation is seldom made use of but for wool intended for samples or tapestry.

The Theory of the Dissolution of Flock.

The reason why from an ingredient, such as the root of madder, perishable colours are produced from dissolved flock, is not difficult to assign. In the first operation of maddering the flock, the red of the madder was fixed in the hair by the preparation of allum and tartar as much as possible, but as it is overloaded with this colour, it is easy to conceive that the superfluous colouring atoms being only applied on those which already filled the pores of this hair, these alone are really retained in the pores, and are cemented by the salts. The hair thus reddened by the madder so as to become almost black, would lose a great deal of the intensity of its colour, if it was boiled in any liquor, was it even common water; but to this water, pearlash is added in equal weight with the flock already dyed, which is to be melted in it; consequently there is a very strong lixivium of fixed alkaline salts made. I have said that very strong alkaline lies destroy the natural texture of almost all animal substances, as also gums and resins; in short, that an alkaline salt is their dissolvent. In the present operation, the lixivium or the pearlash is very concentrated, and very acrid, and consequently in a state to melt the hair, which is an animal substance, which it does very quickly, and with a strong fermentation, which shows itself by the strong and violent elevation of the liquor: consequently it destroys the natural texture of each of these hairs, and the sides of the pores being at the same time broken and reduced to very minute parts, these sides having neither consistence nor spring to retain these salts, and the colouring particles that were sticking to them. Therefore the animal particles of the hair, the colouring parts of the madder, the saline parts of the liquor, and the alkali of the pearlash, are all confounded together, and form a new mixture, which cannot afford a lasting dye, because from these saline parts mixed together there cannot be formed a sufficient quantity of salts capable of crystalization, and producing moleculas, which can resist cold water and the rays of the sun. In short, it could not form a tartar of vitriol, because the alkaline salt is in too great a proportion.

To rouse the deep and overloaded dye of the madder first applied on the flock, and after confounded by the melting of this hair in the mixture already spoken of, putrified urine is added in a considerable quantity; this is a further obstacle to crystalization; consequently wool not prepared by other salts, and dipped in a liquor thus composed, can only be covered by a superficial colour, which finds no prepared pores, or any thing saline in those pores, which may cement the colouring atoms; therefore such a dye must quit its subject on the least effort of what nature soever it be.

But wool prepared by the liquor of tartar and allum, does not take a more lasting colour, in the liquor of the melted flock, than wool not prepared by these salts; for a liquor which abounds with fixed alkaline salts attacks the tartar left of the preceding preparation in the pores of the wool. This tartar changes its nature, and from being hard to dissolve, as it was before, it becomes a soluble tartar, that is, a salt that dissolves very easily in the coldest water.

It may perhaps be objected, that particles of allum remain in the pores of the prepared wool, that from these particles of allum, as well as from a portion of the same salt which is put into the liquor, reddened by the melting of the flock, the alkali of the pearlash must form a tartar of vitriol, which, according to my principles, ought to secure the dye.

To this I answer, that the urine hinders the combination of these two salts, which is necessary for the formation of the tartar of vitriol; if even this hindrance did not exist, the quantity of this salt, which I have named *hard* in another place, could not be sufficient to cement the colour in the pores of the wool, or put them in a state to retain the colouring atoms. Further, the sharpness of the alkaline salts in this liquor, which is capable of entirely dissolving the hair boiled in it, would equally be able to dissolve the wool, were it boiled as the flock was. But yet, though a degree of heat is not given to the liquor, which would be necessary for this total destruction, it is easily conceived, that if the sum of the destroying action is not the same, at least a part exists which, is still sufficient to corrode the sides of the pores of the wool, to enlarge them greatly, and to render them unfit to retain the colouring atoms; to this may be added, that the hair is melted in the liquor, and consequently mixed with the colouring parts of the madder in a great quantity; that these are heterogenious parts, which prevent the immediate contract of the same colouring parts, and that from all these obstacles taken together, the colour must be rendered less durable and less holding than any of the lesser dye. This, experience sufficiently proves, for if a skain of red wool dyed in this manner, be put into boiling water, the colour will be taken off entirely.

Receipt 137th. Scarlet of Archil, and the manner of using it.

ARCHIL is a soft paste, of a deep red, which being simply diluted in hot water affords a number of different shades; there are two kinds, the most common one which is not so good, is generally made in Auvergne, from a lichen or sort of moss, very common on the rocks of that province: it is known under the name of Archil of Auvergne, or Land Archil. The other is a great deal finer and better; it is called the Archil of Herb, or of the Canaries, or Cape Verd Archil; it is prepared in France, England, Holland, and other places.

The workmen who prepare this herb archil, make a secret of the preparation, but the particulars may be found well related in a treatise of *M. Pierre Antoine Micheli*, which bears for title, *Nova Plantarum Genera*, therefore I shall not here give the method of preparing it.

When a dyer wants to assure himself that the

archil will produce a beautiful effect, he must extend a piece of this paste on the back of his hand and let it dry, afterwards washing his hand with cold water. If this spot remains with only a little of its colour discharged, he may udge the archil to be good, and be assured it will succeed.

I shall now give the method of using the prepared archil, but I shall only treat of that of the Canaries, and just mention the difference between it and that of Auvergne. A copper is filled with clear water, and when it begins to be lukewarm, the proper quantity of archil is put in and well stirred: the liquor is afterwards heated almost to boiling, and the wool or stuffs are dipped without any preparation, only keeping those longer in that are to be deeper.

When the archil yields no more colour at this degree of heat, the liquor is made to boil to extract the remainder; but if it is archil of Auvergne, the colours drawn after this manner will be sadder than the first, on account of the boiling of the liquor. The Canary archil, on the contrary, will lose nothing of its brightness, if even the liquor boiled from the beginning. This last, though dearer, yields much more dye, so that there is more profit in making use of it, besides its superiority over the other in beauty and goodness of colour. The natural colour which is drawn both from the one and the other archil, is a fine *gris de-lin*, bordering on the violet. The violet, the pansy, the amaranth, and several like colours are obtained from it, by giving the stuff a ground of blue more or less deep before it is passed through the archil.

It must here be observed, that to have the clear shades of these colours as bright as they ought to be, the wool ought to be sulphured, as was said in the foregoing receipt either before it is dipped in the archil, for the *gris-de-lin*, or before it is dyed blue for the violet, and other like colours.

This way of using archil is the simplest, but the colours that proceed from it are not lasting. It may be imagined that the colours would be better by giving a preparation to the wool previous to its being dyed, as is practised in the great dye, when madder, cochineal, weld, &c. are used; but experience shews the contrary, and I have used the archil on wool boiled in allum and tartar, which did not resist the air more than that which had received no preparation.

There is notwithstanding, a method of using the Canary archil, and giving it almost as much duration as the most part of the ingredients of the good dye; but then its natural colour of *gris-de-lin* is taken off, and it acquires a red or scarlet, or rather a colour known under the name of bastard scarlet. The colours of the kermes or Venetian scarlet, and several other shades that border on the red and the orange, may also be drawn from it. These colours are extracted from the archil by the means of acids, and all those that are thus made may be looked upon as much more lasting than the others, though strictly speaking, they are not of the good dye.

There are two methods of extracting these red colours from the archil. The first is by incorporating some acid in the composition itself that is made use of to reduce this plant to a paste (such as is known to the dyers under the name of archil). I have been assured that it may be made violet and even blue, which probably is done by the mixture of some alkalis, but I must confess I could not succeed in it, although I made above twenty trials for that purpose. I shall now proceed to the second method of extracting from archil a beautiful and pretty lasting red, and which I executed four times with success.

Bastard Scarlet by Archil.

Prepared archil from the Canaries is diluted as usual in warm water, and a small quantity of the common composition for scarlet is added, which is as has been shown in the preceding treatise, a solution of tin in *aqua regia*, weakened with water; this acid clears the liquor immediately and gives it a scarlet colour. The wool or stuff is then to be dipped in this liquor, and left till it has received the shade required. If the colour should not have brightness enough, a little more of the composition must be put in, and pretty near the same method must be followed as in the dying of common scarlet: I tried to make it in two liquors as the scarlet, that is, to boil the stuff with the composition, and a small quantity of archil, and afterwards to finish it with a greater quantity of both, and I succeeded equally; but the operation is longer after this manner, and I have sometimes made as fine a colour in one liquor. Thus the dyer may take his choice of either of these methods.

I cannot exactly fix the quantity of ingredients in this operation. First, as it depends on the shade that is to be given to the stuff. Second, as it is a new process in dying, I have not had sufficient experiments to know with exactness the quantity of archil and composition which ought to be used: the success also depends on the greater or lesser acidity of the composition. In short, this method of dying with archil is so easy, that by making two or three trials in small, more knowledge will be acquired from it than I could teach in a large volume: I must only add, that the more the colour drawn from this ingredient approaches the scarlet, the more lasting it is. I have made a great number of shades from the same archil, and which consequently only differed by the greater or less quantity of the composition, and I always found that the more the archil went from the natural colour, the more lasting it became, so that when I brought it to the shade known by the name of bastard scarlet, it withstood the action of the air and every proof almost as well as that which is commonly made with cochineal or madder.

If too much composition be put in the liquor, the wool will become of an orange colour, and disagreeable. The same thing also happens with cochineal, so that this is not an inconvenience peculiar to this dye ; besides it is easily avoided by proceeding gradually in the addition of the composition, and by putting a small quantity at first.

I have tried the different acids in this scarlet composition, but none succeeded well ; vinegar did not give a sufficient redness to the liquor, and the stuff dyed in it only took a colour of lees of wine, which even was not more lasting in the air than that of the archil in its natural state, and other acids saddened the colour. In short, it appears that (as in scarlet with cochineal) a metallic basis extremely white must be united to the red of the archil, and this basis is the clax of tin. I have repeated the same operation with the archil of Auvergne, but the colours were not near so fine or so good.

Receipt 138th. Red of Brazil or Red-wood.

UNDER the general name of Brazil wood is comprehended that of Fernambouc, St. Martha, Japan, Nicaragua and some others, which shall not here distinguish, since they are all used after the same manner for dying. Some give greater variety of colours than others, or finer; but this often proceeds from the parts of the wood being more or less exposed to the air or that some parts of it may be rotted. The soundest or highest in colour are to be chosen for dying.

All those woods give a tolerable good colour, either used alone, mixt with logwood, or with other colouring ingredients. It will be shewn that, in the false or bastard violet, a little Brazil was added to the logwood; but in the vinous greys, or those which have a cast of the red, a great deal more is used. Sometimes only, a small quantity of galls is put with the Brazil, and it is saddened with copperas; often also with logwood, archil, or some other ingredient, it is added according to the shade, from whence it is not possible to give any fixed rule for this kind of work, on account of the infinite variety of shades which are obtained from these different mixtures.

The natural colour of the Brazil, and for which it is most used, is the false scarlet, which appears fine and bright, but far inferior to the brightness of the cochineal or gum-lacque.

To extract the colour from this wood, the hardest water, such as will not dissolve soap, must be made use of, for river water has not near so good an effect; it must be cut into chips and boiled for three hours; the water is then taken out and put into a large vessel, and fresh well-water put on the wood and boiled again for three hours; this water is added to the first.

This liquor, which is called juice of Brazil, must be old and fermented, and rope like an oily wine, before it is fit for use. To extract a bright red from it, the stuff must be filled with the salts of the common liquor of preparation, but the allum must predominate, for the tartar alone, and also sour water, greatly spoils the beauty of this colour : in short, acids are hurtful to it, and dissolve its red colouring part. Four ounces of allum for each pound of stuff is to be added to the liquor, and only two ounces of tartar, or even less. The wool is to be boiled in it for three hours ; it is then taken out and gently wrung, and thus kept moist for eight days at least, that by the salts being retained it may be sufficiently prepared to receive the dye. To dye with this, one or two pails full of the old juice of Brazil is put into a convenient copper, and well scummed. Dip the stuff which has remained eight or ten days moistened in the preparing liquor, and it must be well worked in it without making the liquor boil too strongly, until it be smoothly and equally dyed. Care must be taken to wring a corner of this stuff now and then, as I have already said, to judge of its colour, for whilst wet, it appears at least three shades deeper than when dry. By this method, which is somewhat tedious, very bright reds are made, perfectly imitating certain colours the English sell under the name of Campeachy scarlets, which by the proof of dyes, are not found to be better than this, only that they seem to have been lightly maddered.

This red, of which I have given the process, and which is no where else described, withstands the weather three or four months in the winter, without losing any of its shade ; on the contrary, it saddens, and seems to acquire a ground, but it does not stand the proof of tartar.

Some dyers of the great dye use Brazil to heighten the red of madder, either to save this root, or make its red more bright than usual. This is done by dipping in a Brazil liquor a stuff, begun with the madder, but this kind of fraudulent dye is expressly forbid by the French regulations, as well as any mixture of the great dye with the lesser, because it can only serve to cheat, and to pass for a fine madder red, a colour which in a few days loses all its brightness along with the shade, which has been drawn from the Brazil, prepared in the common manner.

The first colour extracted from this wood is not of a good dye, probably because it is an indigested sap, and whose colouring particles have not been sufficiently attenuated to be retained and sufficiently fixed in the pores of the wool dyed in it. When these first gross parts of the colour have been carried off, those that remain in small quantity are finer, and mixing themselves to the yellow parts, which are furnished by the pure woody parts, the red resulting from it is more lasting.

By the means of acids, of what kind soever, all the red colour of this wood is carried off or disappears; then the stuff that is dyed by it takes a hind colour, more or less deep in proportion to the time it is kept in the liquor, and this colour is of a very good dye.

It is said that the dyers of Amboise, have a method of binding the Brazil colour in this manner; after their stuffs lightly maddered have been passed through a liquor of weld, and consequently boiled twice in allum and tartar, they put arsenic and pearlash in the juice of Brazil, and it is asserted that this colour then resists the proofs; I tried this process, but it did not succeed.

When a very bright red is required from the Brazil, I know by experience that it is possible to insure the colour drawn from it after such a manner, that, having exposed it thirty days to the rays of the summer's sun, it will not change, but these kind of colours are coffee and chesnut purples.

To make these, I keep the stuff moistened in its liquor in a cellar for fifteen days ; this liquor is prepared as for the reds, of which I have heretofore spoken ; I fill a copper to two-thirds with well water, and the remaining third up with Brazil juice, to which I add about one ounce of Aleppo galls in very fine powder to every pound of stuff, and then boil it one or two hours, as I want the shade to be in deepness : the stuff is aired from time to time, and when it has taken the colour desired, it is well cooled before it is washed. This stuff being brushed, the nap layed and cold pressed, comes out very fine and very smooth.

As to brazil and other subjects for red of the lesser dye, they must all undergo a similar preparation as has been described, and when the red of these subjects are connected with other dyes, you will see it fully described in the receipts of the preceding work.

I shall close this subject of the red, by giving some remarks on the experiments of cochineal liquor.

Zinc dissolved in spirit of nitre changes the red of cochineal to a slaty violet colour.

The salt of lead, used instead of cream of tartar, makes a lilac somewhat faded ; a proof that some portion of lead is joined to the colour of the cochineal.

Vitriolated tartar made with potash and vitriol destroys its red, and there only remains an agath grey.

Bismuth dissolved in spirit of nitre, weakened by an equal part of common water, and poured on the liquor of cochineal, gives the cloth a dove-grey, very beautiful and very bright.

A solution of copper in spirit of nitre not

weakened, gives to the cochineal a dirty crimson.

Cupullated silver a cinnamon colour a little on the brown.

Arsenic added to the liquor of cochineal, gives a brighter cinnamon than the preceding.

Gold dissolved in aqua regia gave a streaked chesnut, which made the cloth appear as if it had been manufactured with wool of different colours.

Mercury dissolved with spirit of nitre, produces pretty near the same effect.

Glauber's salts alone destroys the red, like the vitriolated tartar, and produces like that an agath grey, but not of the good dye: because this salt easily dissolves even in cold water, and besides it calcines in the air.

The fixed salt of urine gives a cinder-grey colour, where not the least tincture of red is perceived, and like the foregoing is not of a good dye, for it is a salt that cannot form a solid cement in the pores of the wool, as it is soluble by the moisture of the air.

CHAP. IV.

Receipt 139*th.* *OF BROWN.*

BROWN is one of the primary or material colours; it is fourth in rank, and it has a great number of variable shades, and is dependent on the power of the corrosive, from the darkest to the lightest shades, let the subjects be of what rank they will, either inclining to blue or yellow, red or black, they must be corroded, before it can be a real colour, otherwise it would be a mixture and it would be no colour in itself, yet there is no colour that has so great a connection with the mixture of colours as the brown, as will be shown in the sequel; it has a variety of subjects as will be shown, and its corrosive powers is copperas; the subjects are so numerous I shall only mention the principle ones. Brown is placed in this rank, because it enters in the composition of a great number of colours, as you may see in the preceding work, in the receipts for browns, &c.; the working is different from others, for commonly no preparation is given to the wool to be dyed brown, and like the blue it is only dipped in hot water. The rinds or barks and roots of the butternut, walnut and hickory, the barks of white oak, of chesnut, of maple, of alder, nut galls and the galls of all oaks, santal, sumac, roudoul or sovie, soot, &c. are used in this dye: butternut bark is the one most in use, and may be ranked as the first, it produces a great variety of shades and if rightly used its colour is permanent, and is one of the greatest colouring substitutes in the Northern States; it is good in all browns mixed with brown and yellow, or brown, red and yellow, as you may see by my former work. As the bark of butternut is so common, and so well explained in the receipts in the preceding, I shall say no more of it here.

Browns are all saddened or darkened with copperas in proportion to the shade required; the other colouring subjects for brown will be noticed hereafter.

CHAP. V.

Receipt 140*th*. *OF BLACK.*

BLACK is the fifth primary or material colour; its principle subject is logwood; it is governed by the power of the corrosive, in which all colours and shades are absorbed and corroded in darkness. The barks, galls, sumac, &c. serve to make a body with the goods for the logwood to act upon, the acid of argal and the alkali, corrects the vitriolic acid, that it receives by the green vitriol or copperas; this vitriolic acid rouses the logwood and gives it a purple brown for which it must be corroded either by acid or alkali, or both. Black has but one shade; that of black or darkness, yet it is denominated as having four, blue black, because the blue is not absorbed; grey black, the pores or bodies are not filled; brown black, for want of a neutral to correct the vitriolic acid and the slightly parts of the logwood; and the coal black, that is, perfectly fine and velvety.

Receipt 141*st.* *FOR BLACK.*

FOR one hundred pounds of cloth, fill your copper with water, then add sixty pounds of logwood chips, thirty pounds of sumac and three pounds of nut galls, or white oak bark as prepared for tanners may be substituted for nut galls; heat and boil well one hour, then run your cloth one and an half hours; then take up and cool, boil again, and run as before; cool, and take two pounds and a half of pearlash, dissolve it in six gallons of warm water, then pulverize one pound and a half of verdigrease, and add one gallon of the pearlash liquor; let it simmer over a moderate fire with often stirring, but not boil; then take thirty pounds of copperas and put with the remaining pearlash liquor, and dissolve it, then add it to the dye, run your cloth one hour, take out and cool; then add the verdigrease solution, run again with the dye boiling, run and air as before; then add three pounds and a half of blue vitriol, run again and you will have a fine black. The verdigrease and blue vitriol, stand as neutrals in this dye; the verdigrease is a mineral of copper, and is much finer than iron, of course has a smaller quantity of earth with it; it is an assistant in saddening, and rouses the logwood: blue or a Roman vitriol, is a mineral of a vitriolic substance, and they both serve to assist and correct the logwood and the vitriolic acid of the copperas, they are both corroding and acid powers in this dye and all other dyes where used.

Receipt 142d. Another form for Black, in which the brittleness and weakness of the goods is completely remedied.

FOR one hundred weight of cloth, fill your boiler with fair water, take twenty pounds of yellow oak bark ground as for tan, or twelve pounds of fustic chips as a substitute for the bark, boil well one hour; then add twenty five pounds of copperas, rake the dye well, then run your cloth one hour, take out, air and return again, and reel with the dye boiling as before; then air and rince your cloth clean, shift your liquor from your boiler, clean well, fill with water and add forty-six pounds of logwood chips, twenty pounds of sumac well dried, and three pounds of nut galls pulverized, or twelve pounds of white oak bark as for tan may be substituted for galls; boil one hour, then run your cloth two hours; then take up and cool, boil again a few minutes, run as before and air; then add three pounds of pearlash or potash dissolved, and two pounds of blue vitriol, boil well, run your cloth two hours with the dye boiling, and your black will be fine and affixed, not superficial and smutty: the black will be soft and velvety. I shall now proceed to give the reasons why blacks are so tender, brittle and smutty; goods are not brittled altogether by the vitriolic acid that the copperas contains, as is the general opinion; first, copperas is made up of three parts, earth, iron and acid, and when applied to the dye of black, according to common form after the vegetable astringents of logwood, sumac, galls, &c.; the earth of the copperas being the lightest part first enters the bodies; the pores and fibres of the wool are closed by the astringents, and the vitriolic acid has not power to force out the earth and gain admission for the iron, because it is so inclosed, and the fibres shut by the astringents that it never enters and preys on the bodies, but remains only superficial. Galls are the most astringent of any vegetable, and when any of these vegetable astringent substances are first used without a preparation, the salts after they are applied, remain superficial and never enter the bodies of the wool; and further it binds all the fibres and bodies of the wool with all the resin and glutinous substance that remains of the animal in the bodies of the wool, and that resin becomes affixed and causes it to crock. The wool is brittle because the earth has entered the pores of the wool, and is bound by the astringents; for that reason the iron and acid remains on the outside and never enters the bodies of the wool, and it cannot be affixed, but remains superficial. It may be asked, why do not blacks lose all theit colour? I answer it is not the affixity, but being loaded with such a mass of colouring substances that the rays of light cannot reflect upon it, and after it has lost half the colour, that it first received, there is a sufficient body to resist the rays of light; for example, take one-tenth part of the colouring ingredients and apply them exactly in the same form as for black, and you will find it will admit the rays of light, and will soon lose all its colour; this shows that the colour is not affixed but superficial: these are the reasons why blacks in general are tender and liable to be smutty; in short not to tire the patience of the reader, I have endeavoured to show the cause of the tenderness and liability of blacks to crock; I will endeavour to give the process (by differently applying the goods and colouring subjects) to prevent their being tender and smutty. By first preparing the goods with copperas and fustic or bark, a portion of the acid of the copperas evaporates, and the earth principally unites with part of the acid and the colouring subject and drives out the colour of this subject, and leaves the astringency; and when the goods are entered, the iron and colouring substance, with a small portion of the acid enters the wool and becomes affixed, and the air drives it into the pores and crystalizes it, that the iron and acid cannot be dissolved by air and water; by rincing in cold water it removes all the useless substance, and part of the earth and acid, and is divested of all the superfluous matter; by emptying the copper you are divested of the whole of the earth, that the copperas contains; now it is prepared for the logwood, sumac, galls, &c.; these astringents take immediate hold of the bodies of these prepared goods, and becomes affixed in all the pores and fibres of the wool. The alkali of pearlash, &c. does not bind the colour, but only corrects and takes off the light and fleety part of the logwood; the verdigrease and blue vitriol, rouses the colouring substances; as acids, they are more so than copperas, and the mineral much finer. The blue vitriol is possessed of a large portion of earth, much more than copperas or verdigrease, and is very astringent and ought never to be used in a preparation, only in the last of a dye after the goods have had a preparation and the fibres of the wool closed by the astringents, then the earth of the vitriol remains superficial and is all dissolved or washed away by water; but as an astringent, it is the most so of all vitriols, it binds the colouring substances, but corrects none.

The cleansing and scouring of the blacks ought to be noticed: the ancient form of cleansing black is, after the loose dye is rinced off, then fill them with fullers earth, it only works through the cloth in the mill, and by this fritting it swallows up all the superficial part of the dye, and rince with clear water: but this is not the best way, after you have rinced off in the mill the loose dye hang the cloth out and dry, and to every hundred pounds of cloth take two pound of hard soap, dissolve it in warm water sufficient to wet them, say twelve gallons, then take three pints of beef gall, mix it with the soap water and sprinkle on the cloth, let it run in the mill fifteen minutes, and then rince well. The soap removes all the loose parts of the dye stuff, and the beef gall makes them all affixed and binds the whole, as it is an astringent. Some erroneously use soap only, but that is the reverse, and leaves the wool too open, like other alkalis.

OBSERVATIONS

ON THE

MIXTURE OF COLOURS,

DYE STUFFS, &c.

CHAPTER VI.

On the mixture of Colours.

ON the five principle colours in this and the preceding work, I have endeavoured to point out the best methods for practice, both in the greater and lesser dye. I shall now endeavour to show the connection these colours have by twos, and by threes, but it will be generally on the great or good dye; it will be needless to have a repetition in this work, as there is in the former work above fifty receipts which give a variety of shades of the lesser dye, and it is well explained in the essay on the properties and effects of dye stuffs & their depending powers.

On mixing Colours three by three.

Blue, yellow and black, produce all dark greens to a black.—Blue, red, and yellow, produce olives, greenish greys and other colours of the same kind; when the mixture contains blue it is usual to begin with that colour.—Blue, red and brown, produce from the darkest to the lightest shades. Blue, red and black, produce a numerous variety of shades. Blue, yellow and brown, produce greens and olives of all kinds.—Blue, brown and black produce olives and greenish greys.—Red, yellow and brown, produce orange, gold colour, burnt cinnamon and snuff colours of all kinds.—Red, yellow and black, produce a colour which resembles a withered leaf.—Lastly, from yellow, brown and black you obtain hair colour, nut brown, &c. Four of these colours may be mixed together, and sometimes five, but this is not usual. It is needless to enlarge upon this subject, I shall only observe that a variety of different shades may be obtained from each of these colours; the design of this enumeration is only to give a general idea of the ingredients that are proper for the production of colours, composed of several others. As it respects the lesser dye of grass green obtained from chymick, it is well explained in the former work ; I shall only mention the process for pea-green, and refer to receipt No. 6, for the preparation of the chymick for the blue: the goods being well scoured are to be allumed, for every twenty pounds weight, two pounds of allum is to be put into a copper with fair water, and the goods boiled gently an hour and a half; whilst this is performing, another copper is got ready, in which fustic chips are put to boil; if there are any to dye pea-green it is best to dye them first, not as practised in some dye-houses, for this great reason, that when several parcels of goods have been through the same liquor, there remains a scurf which the acid extracts, and that is sure to stick to the next parcel that goes in; and if pea-green was the last, the colour would be dulled thereby. The greens (pea-green excepted) are to be turned about ten minutes in the allum liquor after they are dyed, in order to clear them of the stuff, and render the colour brighter. The allum liquor is not to be hotter than that the hand may be borne in. Observe, if the allum was put in (as is customary in some dye-houses) with the fustic, it would retard its working so well; for allum, being an acid, would discharge if used with, as well as prepare for fustic.

OF PURPLE.

IN the first ages of the world this was esteemed the richest of all colours. Purple was the colour of the garments that designated men who were distinguished by their civil and religious stations This beautiful colour was obtained from a shell fish resembling the oyster ; it is taken on the without any other ingredient this fish coast of Palestine ; colours the purple ; it gives a bright and lasting colour to all goods that have received its impression ; this dye stuff comes so highly charged that it has never been much used in any part of Europe or America.

The Grecians found a substitute for purple in a plant called amorgis ; it is probable neither of these will be used in this part of the world, as both are very expensive

OF ORANGE.

THE brightest orange is raised by first colouring the cloth scarlet, and then dipping it in a yellow dye of turmeric and fustic ; it may be obtained by colouring the cloth crimson and then yellow, or first dipping it in a red-wood or madder dye, then in the yellow dye, &c.

Of the mixture of the Colours three by three.

I will again repeat the primitive colours three by three, to impress on the dyers mind what he ought to begin with, and the preparation to govern the dye.

From blue, red, and yellow, the red olives and greenish greys are made, and some other like shades of little use only for spun wool designed for tapestry. It would be a repetition to give the method of using these colours, having sufficiently explained it in the preceding pages.

In the mixture, where blue is a shade, it is usual to begin with it; the stuff is afterwards boiled to give it the other colours, in which it is dipped alternately one after the other; notwithstanding they are sometimes mixed together, and are as good, provided they are colours which require the same preparation; for example, the madder-red and the yellow. As to the cochineal and kermes, they are seldom used in these common colours, but only light colours which have a bloom or vinous hue, and which must be bright and brilliant, and then they are not used in the last liquor, that is, the stuff is only dipped in when it has received the other colours, unless they are to be greyed a little, which is lastly done by passing them through the browning. It is impossible to give any precise rules for this work, and the least practice of these rules will teach more than I could say in many volumes.

Olives are made from blue, red, and brown, from the deepest to the lightest, and by giving a little shade of red, the slated greys, the lavender greys, and such like.

From the blue, the red, and the black, an infinite number of greys of all shades are made, as the sage grey, the pigeon grey, the slate grey, the lead grey, the king's and prince's colour, browner than usual, and a variety of other colours almost innumerable.

Receipt 143d. For Fawn Colour and Silver Grey.

FAWN is a lightish sandy brown being very permanent. For twelve pounds of wool, take half a bushel of walnut husks, put them in the copper of clean water, let them boil one hour; then dip your wool three hours; take up and cool, and add four ounces of crude or red tartar, dip again two hours; take up and cool, and you have a durable colour for silver grey.

Receipt 144th. FOR SILVER GREY.

FOR twenty pounds of cloth or worsted, eight ounces of allum and twelve pounds of fenugreek must boil with the goods half an hour; then take it up, and add one pound of pearlash and eight ounces of Brazil wood; boil them gently with the goods half an hour; rince it and you have a beautiful colour.

From yellow, blue, and brown, are made the greens, goose dung, and olives of all kinds.

From brown, blue, and black, are produced the brown olives, and the green greys.

From the red, yellow, and brown, proceed the orange, gold colour, marigold, feulemort, old carnations, burnt cinnamon, and tobacco of all kinds.

Receipt 145th. For Tobacco or Snuff Colour.

FOR every hundred pounds of wool take twenty pounds of good ground camwood, boil well, run or dip your wool three hours; then have another liquor prepared of eighty pounds of fustic and ten bushels of butternut bark, boil well till the strength is well out; take out the chips and bark; stir or dip the wool six hours; then air, and add ten pounds of copperas and three gallons of sig, immerse again, and keep it in with the dye boiling, till you obtain the shade required This is a most excellent and permanent colour.

I offer these suggestions that workmen may suit themselves in mixing colours. Europeans apprised of our increasing manufactories, attempt to baffle out attempts by imposing on us mixed cloth as fashionable; they are sensible that the younger look to the older nations or the patterns of their garments, and for fashionable colours of their cloths; for this reason the Europeans frequently change or mix their colours to retain our adherence to their markets.

Of Colours which will endure milling.

DEEP blue with all its shades, black, red brown, smoke and snuff browns, cinnamon, crimson, madder-red, pink, purple, claret with red-wood, all browns and drabs. I insert these observations to accommodate those people that would wish to mix any of these colours in the wool for cloth that is to be milled.

CHAP. VII.

A few remarks on Dye Drugs, Woods, and Barks, and Salts preparatory to them

OF INDIGO.

THERE are various qualities, and many dyers fail in their judgment of the indigo; the best is imported from Spanish America, it is generally soft, and will swim on the surface of water, and is called flotong, this is the best kind of indigo for blues, and no other ought ever to be used for saxon greens. French indigo is much harder and in lumps about two inches square; if good, when broken it will appear a fine purple, this will make a fine blue. Carolina indigo will answer for almost all colours, where indigo is used, if it will mix well with oil of vitriol, it will answer for all blues; this kind of indigo is in lumps about the size of French indigo; you may break a lump and find its quality by cutting or scraping it with the edge of a knife, and wetting and rubbing it on the nail: if the colour adheres to the nail it may be pronounced good; but if it appear of a purple, and something mouldy as if the air had passed through it, or puts on a sad, dirty, dull colour, breaks hard and flinty, and is full of small round white specks, it is fit for no use, and will answer no purpose in dying.

OF COCHINEAL.

COCHINEAL is an insect cultivated in South America, it is shipped to Spain, from Spain to England and from thence to Amerca at a high price on account of its accumulated and heavy duties. It is a strong and good dye drug, and will return a handsome profit to the dyer when used in scarlets, pinks and crimsons. That which is good will appear plump and look as though a light sprinkling of flour had been cast upon it. If you keep it dry in a glass bottle, stopped tight, it will remain good many years. There is a kind of cochineal wild and uncultivated, it is small and shrivelled, will make a good colour, but will require three times the weight of the other. Some cochineal is damaged by salt water; this appears of a dirty crimson cast, and is fit for no use.

OF CAMWOOD.

CAMWOOD is with propriety called the best of dye woods; I think it must be a species of saunders; its colour is permanent, and will resist the influence of the air and almost all acids. It is not many years since the use of it was first known in the United States; it is in logs of wood from six inches to a foot through; it splits freely and when good is heavy; on opening it, the first appearance is a bright reddish orange, on being exposed to the air it turns reddish; its smell is pungent. It is much more convenient for the dyer to have it ground, or you must chip it fine; it being a very close wood it will require much boiling: that which is ground, if good will appear of a yellowish red; if you wave it a hot dust will arise, which irritates the nose and the glands of the throat; that which is mild and of a darkish red has been leeched and will produce no good colour.

OF BRAZIL OR RED-WOOD.

RED-WOOD makes fine colours and is useful in many dyes, whether used alone or with logwood; if used with logwood it will produce violet lilac, and is useful in browns where red is required; it is good for a pink, claret, &c.; it is better to use nut galls with it Brazil comes in small sticks; if good, looks bright, of a little yellowish red, smells agreeable, and chips freely. The colour obtained from this wood is not lasting if obtained hastily; the liquor ought to sour, then the colour will be permanent; that which has been damaged by the sea-water or otherwise, afford a dull red chip, and is cohesive and clingy.

OF NICARAGUA WOOD.

NICARAGUA is in sticks of various sizes; these sticks have a number of concaves in them, which have the appearance of art. This wood splits freely, and is of a reddish orange colour, it gives a bright colour and is used much the same as brazil wood, but is preferable in browns, &c.

OF BARWOOD.

SOME have mistaken barwood for camwood, and not knowing the use, both have been condemned. Barwood will dye chocolates and darkish browns; it commonly comes in clefts, it is of a reddish brown, splits freely one way of the grain, the other hard and rough.

Chymical History of Saunders, and its difference from her Red-woods.

RED saunders is a hard, compact. ponderous wood, of a dark blackish red on the outside, and a light red colour within; of no particular smell or taste It is brought from the Coromandel coast and from Golconda. Of the tree we have no certain account. Its principal use is as a colouring drug Those whose business it is to rasp and grind it into powder probably employ certain saline or other additions to improve the colour; whence the remarkable differences in the colour of powdered saunders prepared in different places That of Strasburgh is of the deepest and liveliest red. Some sorts are of a dead dark red, and some of a pale brick red; some incline to purple or violet, and some to brown.

The colour of this wood resides wholly in its resin, and hence is extracted by rectified spirit, whilst water, though it takes up a portion of mucilaginous matter, gains no tinge, or only a slight yellowish one From two ounces of the wood were obtained by spirit of wine three drachms and a half of resinous extract, and afterwards by water, a scruple of mucilage. By applying water at first, I obtained from two ounces two drachms and six grains of a tough mucilaginous extract, which could not easily be reduced to dryness. The remainder still yielded, with spirit, two drachms of resin. The indissoluble matter weighed, in the first case, an ounce and a half and fifteen grains; in the latter, nineteen grains less. Neither the distilled water nor spirit had any remarkable taste or smell.

The red colour of saunders appears to be no other than a concentrated yellow, for by bare dilution it becomes yellow. A grain of the resinous extract, dissolved in an ounce of rectified spirit, tinges it red, but this solution, mixt with a quart of fresh spirit, give only a yellow hue. Hoffman reports that this resin does not give a tincture to any kind of oil. I have tried five oils, those of amber, turpentine, almonds, anniseeds, and lavender. It gave no colour to the two first, but a deep red to the last, and a paler red to the other two.

OF YELLOW WOODS AND DRUGS.

OF these I shall make but few remarks as they have been well explained in the preceding.

Of Weld.—Weld is a plant that may be cultivated among us; it is two seasons coming to maturity and must be cut and cured when in the bloom or blossom, and dried without wet and put up in casks for use: this gives the best and most permanent yellow.

OF FUSTIC.

FUSTIC is the wood or species of mulberry-tree, growing in Jamaica and Brazil, called by *Sir Hans Sloane, Morus Fructu Viridi Ligno Sulphureo Tinctorio.* It is of a deep sulphur yellow colour, which it readily gives out both to water and spirit. The watery decoction dyes prepared woollen of a very durable orange yellow: the colour is imbibed by the cloth in a moderate warmth without boiling.

The fustet or fustel of the French is a yellow wood or root very different from our fustic. It gives a fine orange dye to woollen, but the colour is extremely perishable in the air. The plant grows wild in Italy and Provence, and is cultivated with us in gardens on account of the beauty of its flowers. It is called *Venice sumach, cotinus cotiaria, coccigria; cotinus matthioli, C. B.*

As to yellow wood, green wood, turmeric, &c. I shall make no further remarks.

Of Logwood as a colouring drug.

LOGWOOD or Campeachy wood (*Lignum Brazilo-simile, cæruleo tingens, J B.* is the wood of a low prickly tree, which grows plentifully about Campeachy or the Bay of Honduras, and has of late been introduced into some of the warmer of the British plantations, particularly Jamaica. It is a native of the low marshy places. The wood comes over in pretty large logs, cleared from the bark. It is very hard, compact, heavy, and of a red colour.

Logwood gives out its colour both to watery and spirituous menstrua, but not readily to either without boiling; it requires to be chopped fine or ground, and damped with water a month or two before use, when it gives more colour and is easier extracted. Rectified spirit extracts the colour more easily, and from a larger proportion of the wood than water does.

The tinctures both in water and in spirit are of a fine red, with an admixture, particularly in the watery one, of a violet or purple. Volatile alkaline salts or spirits incline the colour more to purple. The vegetable and nitrous acids render it pale, the vitriolic and marine acids deepen it.

The watery decoction, wrote with on paper, loses its redness in a few days and becomes wholly violet. This colour it communicates also to woollen cloth previously prepared by boiling with a solution of allum and tartar. The dye is beautiful, but very perishable. It is often used by the dyers as an ingredient in compound colour, for procuring certain shades which are not easily hit by other materials.

With chalybeate solutions it strikes a black. Hence it is employed in conjunction with those liquors for staining wood black for picture frames, &c. and with the addition of galls for dying cloth and hats black. The black dyes in which this wood is an ingredient, have a particular lustre and softness, far beyond those made with vitriol and galls alone. The beauty however which it here imparts is not permanent, any more than its own natural violet dye.

On the same principle it improves also the lustre and blackness of writing ink. Ink made with vitriol and galls does not attain to its full blackness, till after it has lain some time upon the paper. A due addition of logwood renders it of a deep black as its flows from the pen especially when vinegar or white wine is used for the menstruum

Dococtions and extracts made from logwood have an agreeable sweetish taste, followed by a slight astringency. They have lately been introduced into medicine, and given with success in cases where mild restringents are required. They often tinge the stools, and sometimes the urine of a red colour.

Of Copperas or Green Vitriol.—Copperas is an extract of iron corroded by acids, commonly old rusty iron and vinegar, this is the reason of its containing so great a proportion of earth, and congeals into a salt; as a substitute for copperas take of the filings of iron, and put them in vinegar and let it stand a month, you will have a much purer darkening substance. The best copperas is the brown, or that which appears to be mouldy, it is the cream of the mineral; the deep green will make the brightest blues and browns, but is not so strong as the other. and will not make so good a black; that of a pale green colour is worth but little. The iron is corroded with different acids, as oil of vitriol, &c. and will answer no purpose in dying: copperas ought to be kept in a celler where it is not very damp nor open that the acid may evaporate.

Of Nut Galls —Nut galls are of great use in dying black and greys; galls are the basis in the ground preparatory to all cotton dying, except blue, the astringency of the galls becomes affixed on the body of the cotton, and the colouring substance immediately adheres to it. The galls come from Aleppo and Smyrna; the Aleppo galls are generally the best, they come highly charged, and are not so much used as they ought to be in dying: some barks may be substituted, as will be mentioned; our oak galls gathered and cured in their season will answer nearly the same purpose, and it is wished that those who have oak groves would gather them, that they may be brought into use.

Of Sumac.—Sumac is a crooked shrub with spreading branches of different heights, and grows spontaneously in many parts of the country. It is used in three different ways; when the wood is used only, the bark and sap must be shaved off, as there is a glutinous balsam in the sap and bark which will adhere to the cloth and will form a resin, that will have a bad effect on the goods; the other methods are to cut the sprouts and branches with the bobs or berries of one season's growth, make or cure them as you would hay, without wet, and put them up for use; in the third method the process is the same in cutting and curing, it is then conveyed to the sumac factory, where it is manufactured and put in casks; this is the best for common use. It never ought to be used green, on account of the gum, which evaporates or disappears in the curing and manufacturing. The wood is useful in drabs; and the manufactured in blacks, browns, &c

OF BARKS IN GENERAL.

BUTTERNUT bark is the best colouring substance in North America, it will give a variety of shades, and if used right its colour is durable; it is good in many browns but not in black. It is best to use it when green, through the autumn and winter seasons; the wood ought to be cut in the last of November, for the winter's use, and housed, and the bark shaved off as you want to use it; in using it, it should be cut fine, put it in the boiler and put a fire under it the day before you begin your colouring, that the liquor may be warm; immerse the goods when it is as hot as you can bear the hand; never suffer it to boil, and the colour will be permanent, but if it boils the colour will not be so bright, the shade will be different and the colour will not be lasting. In the Spring when the bark will slip, peel the bark from the trees, for the Summer's use, and house it immediately after peeling; never cure it in the sun; after it is dry it may be boiled, yet the colour will not be so lasting; by using it green and dry, boiling and not boiling, and by cutting it at different seasons of the year, you may produce a number of shades, from a dark smoke to an orange and flesh colour. It is good in smokes, olives and snuff colours

Of Yellow Oak bark.—This bark gives a lasting colour and is good green or dry, but better if prepared as for tanners, or rasped and ground; it is excellent in black, very good in olives, and is a clean substance.

Of Walnut or Hickory bark —This is much the same as the oak bark, but its colour is brighter, the dye of this bark is durable and will answer the same purpose as the oak: the rind or husks of the nut are most excellent.

Of White Oak bark.—This is a most excellent bark on account of its astringency, its colour is lasting and may be used in any dye that galls are used in as a substitute after it is dry and ground as for tanners.

Of Alder bark.—This bark is good and its impression is durable; it is good in black and almost all dark colours; by filling up the ground of the colour it leaves it bright.

Of Hemlock bark.—Hemlock is a very good bark for colouring, the rap should be taken off; it is good green or dry; in light browns it gives a colour of a reddish cast.

Of Yellow Birch, White Ash and Sassafras barks.—These are good in light browns and ash colours, if used right; the colours will be clear and beautiful, and they will leave the cloth soft and nice.

Of Chesnut, Maple and White Birch barks.—These produce beautiful browns very much alike; they answer in greys, but the colour soon fades. It is to be observed that bark of the roots and the rinds of the nuts, give much more colour than the barks of the bodies and may be used the same as their respective trees.

To preserve Dye-Stuffs from injury.

STRICT attention ought to be paid to this branch of business, as some dye-stuff will loose all its valuable qualities, and some by collecting filth and dirt become useless. Wood in the stick of all kinds ought to be kept in a dry celler, raised from the ground and kept from dirt; all yellow wood when chipped or ground ought to be put in casks and excluded from the air. Camwood and logwood whether chipped or ground ought to stand open in casks, and be kept clean as it improves by the air in a dry celler; copperas may be treated the same way. Madder, woad, indigo, and all aleotics should be kept in a celler and excluded the air, as it destroys all their active volatile substances. All preparatory salts and colouring drugs ought to have close boxes, to keep out the dirt and air. All liquid substances must be put in glass bottles, stopped close with glass or wax stoppers. Galls and all rinds and barks require to be kept in some dry place, or they will be liable to mould and mildew, which will destroy all their colouring substances, &c.

The Cultivation of Teasles.

TEASLES are the most useful and necessary to dress fine cloth well, and without them cloth cannot be well napped, or a good pile or body raised. Among common cloth-dressers they are but little used or their value known, I may say they are not used the tenth part of what they ought to be. This plant is very productive both in burs and seeds, and is easily cultivated; the seeds ought to be sown or planted early in the spring; they are two seasons growing to maturity; when the plants becomes large enough to transplant, set them in moist rich ground, about eighteen inches apart, hoe them, and keep them clear from weeds; during the summer they will have fifteen or twenty long rough leaves; before winter, before it freezes hard, cover the plants with brush, and spread straw over them, as soon as winter breaks take off the brush and straw, and when the weeds arise, hoe them twice or thrice, by June they will begin to stalk and branch out in various branches; from the stalk comes out long leaves within six or eight inches of each other, and form around the stalk, resembling a dish with two long handles, and standing erect; in this bowl or bason spring two other stalks; it supplies itself with water by rains and dews in this bowl; upon the end of each branch is a bur, some one and a half inches diameter, and four inches long, so in different grades, some not larger than the end of the finger: a plant frequently produces one hundred and fifty burs, of which fifty will be fit for use; they rise from three to four feet; the bur is curiously set, resembling the honey-comb, with very sharp points, hooking towards the stalks. After the blossoms have fallen off, is the time to cut them, within about six inches of the stalks, dry them well, and keep them from wet, as the water will ruin them after they are cut and laid down; the bur sheds its water while on the stalk: by this method the factories and cloth-dressers may supply themselves with the most useful implements for dressing cloth: the mode of using them is well known, and how they ought to be used will be explained in the sequel. It is but a short time since they have become a matter of note and speculation among us. I know of a man, who raised one crop of teasles on half an acre of ground, which he sold at wholesale to the dealers for *Four Hundred Dollars*: now my friends use economy, save your money and raise your own teasles, and you will have them when you want them; if you once get in the use of them and have any desire to have your work look well, you will never be without them; they are nature's cards, and nature has provided more than we can by art.

CHAP VIII.

Of Sorting Wool.

THIS is an important branch in manufacturing woollen cloth, as there are more than one hundred different qualities of wool: every fleece ought to be divided in four parcels, that on the neck and fore shoulders of the sheep, is the finest; that on the back and partly on the sides, the next; on the belly, the third; on the legs and hinder parts, the fourth: in this form, the wool ought to be assorted, from all species of sheep. Among the different species of sheep, are the merino, full and part blood, the English, the common, the Vienna, the cape sheep, &c.; from these sheep we have almost innumerable qualities of wool, which ought to be well assorted, and kept separate for their several uses. The best wool for superfine broad cloth, is the thick set, fine and curly wool, and is the worst to work; the second pick of the full blood is proper to match with the first pick of the halfblood; following this rule, except that what is on the legs and rump of the sheep, which is only fit for listing, carpeting and coarse cloth. There is another quality of wool, long, silky and open, this ought to be combed into worsted; from this proceeds a variety of qualities, that may be wrought into cloth for light weaving: there is still another very coarse and hairy wool, this ought to be wrought into cloth, for bear-skin, lion-skin and baises. Lamb's wool should be made into cloth for flannels, of various qualities, it will be much whiter, will felt better in the mill and nap much easier. It requires strict attention and good judgment to assort wool; on the assorter depends all the beauty or ground of manufacturing cloth; by mixing coarse and fine together you have no distinction in the quality; one pound of coarse wool is enough to destroy the quality of ten pounds of fine, &c.

Of Scouring or Washing of Wool.

This is another very important branch, and very much neglected for three reasons: first, the wool in its natural state is possessed not only with the animal oil, but a sort of gum which preserves the wool on the animal, it keeps out the inclemency of the weather, the heat of the sun, &c. If not divested of this before use it gums and gluts the wool and cards, it forms with the oil that is applied for carding, a sticking glutinous gum which destroys the active life of the wool; it will spin slubby and you cannot have good yarn. Second, if the wool is to be coloured it is very injurious to many dyes, and it requires strict attention from the dyer, as it will soon overset the blue dye, and that is one thing why so many fail in their blues. And thirdly, to admit it never injures any dye; there is another objection, if the wool is coloured with any part of the gum and animal oil, it forms a sort of resin that becomes affixed in the bodies and fibres of the wool by the hot liquor, and never can be removed; and the colour will remain only superficial. From this the dyers, and manufacturers may learn the cause of their cloths crocking.

There are different modes of scouring wool in practice, I shall describe all those worthy of notice; but first point out the way I conceive to be the best, and the mode generally practised in Europe, and in the best manufactories in the United States of America.

In the first place you ought to provide yourself with a boiler that will hold three barrels, it is better to have it of cast iron, as the alkali and animal oil will corrode the copper; have this set near your stream of running water, or your large cistern that will contain two hundred hogsheads of water, for the convenience of rincing; reference ought to be had for convenience of working and heating, and further they both ought to be set with a roof over the sap and open to the sides, that the air may pass through; otherwise the volatile substance of the urine will nearly take the breath, let it stand adjoining the wash-house, or rincing place; then have a wash-box made four feet deep and four feet square, with a sieve or strainer about one foot from the bottom, with a gate or sluice way to take the water out at pleasure. Some use a basket for rincing this is not as good or convenient. Have your box placed so you may easily let the water into it, have another box like a sieve at the bottom to cast the wool in, to drain after rincing. If you wish to make despatch in drying, have a press with a screw to squeeze the water all out and spread it out immediately to dry. The scouring of wool is properly the care of the dyers, let it be for what colour it will, the filth and natural oil of the wool ought to be extracted and it divested of all the gum. The natural oil which adheres to the wool preserves it in the warehouses and also from moths. The process is as follows, to eighteen pails full of water put six pails full of fermented sig or urine, mix them together in the boiler, heat as hot as you can bear the hand in it without scalding; take twenty pounds of wool stir this gently to and fro with sticks for that purpose about fifteen minutes, keeping the heat the same; take it up in a basket, squeeze the liquor from the wool into the boiler, then cast it while warm into the wash-box, set the cold water to it, stir it backward and forward with sticks so as to keep the wool open; then drain off this water, fill the box again with fair water, stir as before till the wool is all open and clean; then with a pole take the wool out and cast it in the other box to drain; while this is rincing another draught may be put in the boiler and thus proceed till the whole is scoured: as the liquor wastes fresh is to be added of one part sig and two parts water, but if the urine is strong and old you may add three parts water. A man will in this way generally scour a bale of wool in a day, if it weighs two hundred and fifty pounds in the fat; it generally less looses sixty pounds in scouring, but the diminution of weight varies in proportion to the wool being more or scoured, and in proportion to the more or less fat contained therein; too much attention cannot be paid to scouring as it is better disposed for the reception of the dyes. This is the best method in scouring and is followed in the manufactory of Audley in Normandy, where cloths are beautifully manufactured.

A solution of soap and water cleans the wool of the filth and oil; there is one objection to this, it felts the wool. Another method comes nearest to the urine, to twenty-four pails full of water take four ounces pearlash and two ounces oil of vitriol, the vitriol neutralizes the pearlash and gives life to the wool and leaves it clean; following the same process in cleaning as before mentioned, the vitriol is a mineral oil, and forming a solution with the vegetable alkali, of these it forms a connection near in substance to the animal alkali of fermented urine, &c.

Of Manufacturing Cloth.

BUT little remains to be said on this subject, more than what is given in the preceding work. After the wool is scoured and dyed, have it looked over, take from it all the burs and dirt, and clip all the dead ends; to sixteen pounds of wool take two pounds of sweet or good sperm oil; then pick with the picker or hand, to mix the oil with the wool to leave it open; then card it into rolls with a machine, or by hand cards, your cards require to be in fineness to your wool; from thence have it spun into yarn, the waff or filling slack twisted; then weave it into cloth; have it sleighed according to the fineness of the yarn, and closed to make it squire as much waff and warp. Be cautious in having good list yarn, and make a good list two inches wide for broad cloth: this list is of no use only to the millman, dyer and finisher, and cloth that is to be milled cannot be handsomely finished without the list; after the cloths are wove, if they are not ready to mill, they must be overhauled every week or ten days, to give them air and keep them from mildew; the oil and size collects dampness and causes a heat or fermentation which will mildew without air, and destroy the life of the wool; when the cloth is wove have it burled or picked of all the knots, burs and doublers carefully, then it will be fit for the mill.

Of Milling Cloth.

OF the fulling mill.—There are various forms in use, and the most of them badly constructed, yet I shall give no form, but let every millman follow his own choice; I will only remark that the falling mill rightly constructed, makes the firmest and best cloth, and is the most difficult mill to tend; the crank mills are the best to scour and wash, are less liable to damage and are easier tended. On the whole, the crank mill ought to have the preference. There are different forms of milling and scouring, and some of them are so bad that the millman ought to be brought to the seat of justice and prosecuted for fraud, or barred the privilege of milling. They will full their cloths in lies, because this method is cheaper than soap: this is a pernicious way of doing business; the cloth will be rough, brittle and will not do half the service, as if fulled in soap; the lie will start the grease; he only saves to himself a few cents while he robs the community of many dollars. Some full the cloth in the grease, till it is sufficiently milled; this is a bad practice, it will leave the cloth loose, and it does not uniformly unite in felting; you cannot have firm, well milled cloths in this way, although it will appear thick. Some leave grease in cloths after they are milled; this is a piece of insufferable deceit and slovenness; when in the cold air, such cloths will appear to be thick and firm, when warm they will be limsy and emit a fœtid nasty smell; you cannot make a bright colour on them; they will smut, and never can be finished handsome, will always be catching dirt, and will not do half the service as when cleansed from the grease.

I shall now give the mode I practise, and the general mode practised in England and France. The stock of cloth ought to be in proportion to your mill, and the mill so constructed as to turn the cloth gradually, every time the hammer fetches up to the stock For the first milling or scouring the filth and grease out of the cloth, to fifty yards of broad cloth or eighty pounds weight, take two pounds of pearlash, dissolve in one gallon of warm rain or river water; then take eight gallons of well fermented urine, mix it together, sprinkle it carefully and evenly over the cloth till the liquor is all on, then lay it in the mill, let run one hour, take out, handle over and speedily lay it in again, let it run one and an half hours; take it out and stretch the cloth all over; lay it in again, run till it forms in a proper body for milling; then turn into the mill gradually five or six pails full of warm water, as warm as you can bear the hand in; when it is all in a lather, let the cold water run on the cloth, till all the sig, filth and grease is washed out: if the cloth twists and binds up, so that it does not run regular, hand over, lay it in again and rince till clean; then take it out on a scray, hang it out to dry; when dry, take it to the burling board, look the stock of cloth all over, pick all the knots, burs, and cotton or linen specks, that remains in the cloth of the second burling; at this time after scouring may be seen all the defects, that will be injurious in finishing, as no burling ever ought to be done after the milling is finished: this is the first milling or scouring it, and divesting it of the filth and grease.

Another method about as good for scouring.

TAKE for a stock as before, eight gallons of good soft soap, eight gallons of hot water, and eight gallons of sig, mix them together; sprinkle it over the cloth, when as warm as you can bear the hand, sufficient to wet the cloth, let run in the mill, till all has received the liquor equally, say ten minutes; take out, hand over, double up close and let lay eight or ten hours; then lay it in the mill, run one hour, and manage as in the preceding, and it will divest the cloths of their filth, grease, &c.: when it is dry and burled, it is ready for the second milling. Take for a stock as before described, white hard soap as made at Roxbury without rosin, as the rosin is injurious to the cloth; it gluts and hardens the wool, that it will not appear fine. Take of white soap, six pounds shaved up fine, put in a tub, add seven gallons of hot water, (but not boiling), stir till the soap is all dissolved;

when it is as warm as you can bear the hand, sprinkle it carefully over the cloth by little and little ; lay it in the mill, let it run one hour ; if not wet enough add a little more soap, but be cautious and not have it too wet as it retards the milling and the cloth will not be as firm : have it so wet that you may easily wring out the soap with the thumb and finger ; as it dries and requires soap, add more ; frequently handling over and stretching the cloth, that it may not grow or adhere ; have your eye at the mill, handle over whenever it does not turn well, stretch once in an hour and a half or two hours, and add soap as it is wanted, till all the soap that is prepared is on if required : manage in this manner till you have brought your cloth to a right thickness and it is well milled, or to the length and breadth required. When it is milled to your liking pour a few pails of warm water gently on the cloths, then rince with cold water till all the soap is extracted and the water runs clear and clean from the mill and cloth ; take it out, stretch and lay it smooth : when it is ready for dressing or finishing.

Some use soft soap for milling, but this is a bad practice, as it is too sharp and fiery, and raises the wool too much ; the cloth will be loose and spungy ; the white hard soap is the reverse, it will make the cloth firm, use as much as it will bear and the cloth will be much better and firmer.

Of Finishing Cloth.

NOT much more remains to be said, than what has been said in the former work ; there are various forms in practice, the same may be said with respect to tools and machinery. Let every workman fix on his own form ; but this much may be said,the beauty of the cloth much depends after it is well milled in raising the nap, and that ought to be done with teasels with the cloth wet. It ought for a superfine cloth to have three good nappings, so as to have the pile cover the thread every time after shearing ; have it sheared even and close twice ; every time you raise on the face side, always raise the nap one way of the cloth, that is leading toward the mark ; when it is sheared and raised the third or last time with teasels it is ready for dying. If not dyed in the wool, all the pile should be raised before the cloth is dyed, as colouring brittles the wool and you can never get a good pile after it is dyed ; when dyed and cleaned from the dye stuff, lay the nap with good limber jacks out of warm water straight and smooth, or with a gig as a substitute for jacks, teasels may be used in a gig also ; then stretch it on the bars straight and smooth, and lay the nap with a brush when wet, then sheer again twice or three times on the face as it requires ; observe never to shear the lists heading and footings, shear once on the back side, look it over and see it is free from specks and defects ; then brush it thoroughly with a brush and sand board, or emery board with it a little damped, roll it hard on a roller, let it remain six hours then fold for the press. If fine cloth put it in good smooth press papers and press cold, screw it very hard; if coarse, press hot and do not screw hard. It is best for a factory to have plates of cast iron about three-eighths of an inch thick to place between each draft, have them made in size to the papers; put between each draft half the size of the papers, heat these plates in a stove for that purpose, let it remain in the press twenty-four hours, then shift the fold; press as before, take it out of the papers and pack fit for market.

Of Sulphuring and Whitening Woollen Cloth.

A TIGHT convenient room is necessary for this purpose, it should be prepared with shutters or scupper holes which may be thrown open when necessary; and drive tanter hooks in the joyce within six or eight inches of each other; for every hundred weight of woollen cloth take six pounds of sulphur, have a number of chafing-dishes or other vessels for that purpose, place them at an equal distance from each other on the floor, put about half a pound of sulphur in each vessel; then have your goods prepared, wet evenly but not so as to drip, with weak soap-suds of white hard soap, then hang it by the lists straight and smooth on the hooks, with one edge hanging down and the spaces between each piece three inches. When thus prepared sprinkle ashes on the sulphur and set fire to it, shut the room tight for six or eight hours, then throw open the shutters or scupper holes to let the sulphureous vapour blow off, for was any person to enter such a room before it is ventulated he would be in danger of suffocating: by this procedure woollen cloth may be rendered as white as India shirting. I will give a few reasons for this effect, the sulphur is a mineral possessed of a great share of acid, and the acid evaporating by the heat seizes immediately on the body of the wool and makes it uniform by adding to those parts that have not sufficient life and taking from those that have too much, and by uniting in all the body of the wool equally when it enters it immediately drives out the alkali of soap, and all the glutinous gum of the animal; as the alkali and acid form no connection, and the acid will corrode the alkali; it is so powerful it will remove all dirt, spots and defects in the cloth. Wool may by whitened or st ved in the same manner, by preparing perches to suspend the wool loose, and it is wished it was put in general practice, as it divests it of all the crusty dead gum which retards the dying by glutting the fibres of the wool, and when it is thus stoved it divests the wool of all its dead substance and gives it a uniform life: the wool has equal life in all the bodies if you divest it of this dead gum which is not equal and uniform;

it is not soluble in water although it may be removed by a preparation of the alkali, as the alkali will dissolve the gum, but if too powerful will destroy the bodies and animal life of the wool, instead of giving life; the acid if too powerful will have near the same effect; but by applying them in this weak and mild way they neutralize each other, and for the same reason it may be used by a solution of the same qualities and avoid the smell of brimstone after this proportion, to every thirty gallons of water take one pound of white hard soap, or two ounces of pearlash heat the water boiling hot; then add four ounces of oil of vitriol, run your goods thirty minutes and rince clean in the mill. Another method of solution for whitening and cleaning woollen goods take of the compound as for Prussian blue and green, only add double the quantity of vitriol you do for green; to sixteen pounds take two tea-spoons full of compound, add warm water near scalding hot, mix it well with the water run your cloth one hour, if it does not blue your cloth too much you may add a little more, observe not to blue it so that it is hardly perceptable. This is the best method for flannels and all other white woollen goods that are to be worn white, as it remains white much longer, and does not yellow as the stoved. The reason of this is the fibres are a little filled with the colouring atom; while on the other hand, the bodies are all open and exposed to the vapours of the air and becomes affixed the same as on the animal, and are not soluble by water, but must be removed the same as at first, &c.

To know when Cloth has been well Milled, Finished and Dyed.

WHEN cloth has been well milled and finished in a proper manner it will be soft and firm; being shorn even, it will present you a short thick nap which lies smooth in one regular direction; by drawing the hand the way the nap inclines it will feel sleek and smooth; move the hand the reverse the nap will feel rough and prickly: if the cloth will bear this inspection, you may conclude the workman has done his duty. The workmanship on cloth, that is designed for handsome dressing may be discovered by the eye; if it is pressed stiff like buckram, if the nap be irregular and the face of the cloth be rough, the workman has not performed his duty, but has endeavoured to hide his failure by the press. The press on thick cloth is of no importance; cloth should be so dressed as to wear as neatly without as with pressing: the only reason that thick cloths are pressed is to settle the bodies of the wool, and make the threads uniformly smooth and firm, compact and finished. However, if the cloth has not been regularly manufactured before it is delivered to the dyer, millman and finisher, it will be beyond their power to finish it neatly. Whoever will inspect cloth in conformity to the foregoing directions may easily know whether the workman finisher has performed or neglected his duty.

Did the people of this country thus inspect their cloths, unfaithful and ignorant cloth dressers would not be employed; while the well informed, and faithful workman (it must be acknowledged we have some as good and able workmen as in any country progressing rapidly in the improvements of useful arts,) would be enabled to do business upon a more extensive scale, than has been yet attempted in America.

If cloths were manufactured and dressed as well as our wool will admit, gentlemen in general would prefer the productions of their own country to those of Europe: but greatly to our injury, cloths of this country too generally have not been properly treated in dying and dressing: one reason is, because many who pretend to be workmen are entirely ignorant of colours, their combinations, and the physical qualities of dye stuff; another reason that may be rendered for this imposition is, because many attempt to dress cloth before they are acquainted with the business, and consequently never acquire a suitable knowledge of it. It would greatly promote the interests of the nation as well as that of individuals were no person to attempt the dying and finishing of cloths, until he had acquired a suitable information by instruction and experience: gentlemen of literary acquirements who have turned their attention to chymical analysis, acknowledge that the art of dying is as difficult as it is useful.

A great proportion of the people being unacquainted with the clothiers and dyer's art, have been satisfied with the workmen they employ, though their goods have suffered through the ignorance or fraud of the dyer, millman and finisher. If the goods present a flashy and fanciful colour, and come stiff from the press, many people suppose they are well dressed; but the stiffness which the cloth has acquired from a hot and close press is designed merely to conceal the faults of the finisher. The populace will find on wearing such goods, that the colour will soon fade, and the cloth soon become rough and appear coarse, whereas if the cloth had been well coloured and dressed, it would have worn smooth as long as the garment remained whole and decent.

For general information it may be necessary to point out some further directions that any person, on viewing a piece of cloth may determine whether it be well coloured or not.

Of Colours.—Some reflect a beautiful lustre from the extremities of the nap, that is raised on cloths; others present a beautiful body from the grains of the cloth, but afford no lustre; those which afford a lustre or reflect the rays of light that incidentally fall upon them, are the deep blues, all greens, black, red browns, purple, cinnamons, clarets, smoke, snuff and olive browns; these are full colours; if well dyed, by casting the eye towards the light level with the cloth, the hearts of wool that rise up on it will appear bright and lively, as if the rays of light shone through them: those colours which by this experiment appear faint and languid, you may determine have not received their complement of dye stuff and are not well coloured. Scarlet affords no lustre, but if well dyed the body of the cloth will look glaring, bearing slightly on the orange; crimson presents no lustre, but if well coloured gives a beautiful body: some reds produces a lustre and glare full of the blaze. There are many shades of different colours which give no lustre, yet they appear clear and bright.

It is necessary that the dye should equally penetrate the pores of the wool, then the cloth will with few exceptions as to colour, if well dressed appear handsome; but if the cloth has not well received the dye, or it appears daubed, it will discover the fraud or ignorance of the dyer; but if it be poorly finished, however good the colour, the cloth will never afford even a decent appearance.

END OF THE APPENDIX.

THE

DYER'S COMPANION.

PART II.

Containing Many Useful Receipts.

1. *To Jack or harden Leather for Horseman's Caps, Holsters, &c.*

I HAVE found by experience, that saddle leather is the best for caps and holsters. In this case, let the cap, &c. be perfectly dry; and on the block when jacked; take melted rozin, as hot as is convenient, rub it on with a small swab, then pass the cap back and forth through a light blaze, and hold it to the fire till it strikes in; repeat it a second time. It is a repellant to water, and keeps the work in its place. For leather that has not been oiled, add to three ounces of rozin, one ounce of bees-wax, and half an ounce of tallow.

2d. *To make Varnish for Leather.*

TAKE three ounces of gum shellack made fine, and one ounce and a half of Venice turpentine put them into one pint of double rectified spirits of wine, place the bottle in hot sand or water for six hours, shake it often, and apply it with a soft brush or the fingers when blood warm. Repeat it three or four times in the course of twelve hours. If you wish it black for boots or shoes, add half an ounce of ivy black &c.

3d. *To make Liquid Blacking for Boots and Shoes.*

TAKE one ounce of oil of vitriol, one ounce sweet oil, three ounces of copperas, three ounces of molasses, mix them together, let it stand one hour; then add one pint of vinegar shake them well together and it will be fit for use.

4th. *To prepare Feathers, Fur and Hair, to receive Red, Yellow or Green.*

THIS preparation is necessary as the oil must be extracted previous to colouring. For one ounce of feathers, take one quart of water, add to it one gill of sour wheat bran water, one ounce of cream of tartar, and half an ounce of allum; simmer this together; then after the feathers are washed and rinced, put them in, let it stand twelve hours, keeping the liquor hot.

N. B. White only will receive the above colours.

5th. To Colour Feathers, Fur, &c. Red.

TAKE half an ounce of cochineal made fine, mix it with an ounce and an half of cream of tartar to one quart of water; when simmering hot, add a tea-spoon-full, let it stand ten minutes, then put it in the feathers, and so on each ten minutes, until exhausted. In all colouring, the dye must not be crowded, and soft water must be used. After the whole of the colouring is in, let it stand fifteen minutes, then rince them in clear water; whilst in the dye, five or six drops of aquafortis may not be amiss as it sets the colour more on the scarlet.

6th. To dye Brussels, Red.

TAKE one ounce of Brazil wood ground, half an ounce of allum, quarter ounce of vermillion, and one pint of vinegar, boil well, put in the brussels when hot and keep them in till cool, and you will obtain the colour required.

7th. To Colour Feathers, Fur, Hair, and Woollen or Silk, Blue, of any shade.

NO preparation is necessary except washing and rincing. To eight ounces of oil of vitriol, add one ounce of indigo made fine, a tea-spoonfull of each six or eight minutes, shake it often; it must stand two or three days before it is fit for use; indeed the longer it stands the better: one tea-spoonfull of this to one quart of water, when hot as is convenient for flesh to bear, make an azure blue; by adding or diminishing, any shade is produced. It is not recommended for woollen, except for women's light wear, stockings, &c. as the colour is not very durable on the wool. Those light articles being easily re-coloured, it will be found the most convenient and expeditious method of colouring, as ten or fifteen minutes is sufficient for any of the above articles to colour. It is also very useful to revive old dye that has decayed; also, a few drops put into rincing water for silk, stockings, &c. gives the primitive clearness. I am sure, if the use of this was known, that scarce a family would be found without a phial of it in their house; when cold let it be stoped tight with a glass or wax stopper.

8th. For Blue on Brussels.

TAKE one ounce of good indigo, and one ounce of biss, a small nub of allum the size of a hazlenut, one quart of gum water, simmer them all together and dip the brussels when hot; you may substitute one quarter of a pound of gum arabic dissolved in one quart of hot water in lieu of gum water, let them lay in the dye two hours; then take them out, clap them well with the hands, that in dying you may imbibe the colour, hang them up to dry; if different shades are required you may change the order of the dyes, always using gum water or gum arabic dissolved as before; for black, use logwood, nutgalls, copperas, &c. For purple use lake and indigo. For carnation colours, use vermillion and smalt. For yellow, use berries, saffron and tartar, all mixed and dissolved in gum water; use your judgment try and see.

9th. To Colour Feathers, &c. Yellow and Green.

TAKE two pounds of fustick, chip it fine, boil it in two gallons of water four hours, keeping the quantity of water; then take out the chips, and add one ounce of curkemy root, and an ounce of allum; boil the two gallons to two quarts, let the feathers lie in the dye one hour to make them green; add two tea-spoonfuls of the oil of vitriol and indigo. They require to be only rinced after colouring.

10th. For Green on Brussels and Feathers.

TAKE one ounce of verdigrease, one ounce of bees-wax, one ounce of tartar, one gill of vinegar, one quart of gum water or four ounces of gum arabic dissolved in water; mix them all together and heat them, then take the brussels and feathers and dip them in hot water, then in the dye, clap them with the hand, let them lie two hours and hang them to dry.

11th. For Light Green on Woollen.

TAKE of the juice of the herb called horsetail to which add one ounce copperas, one ounce of verdigrease, and half an ounce of allum, heat it hot and handle till your colour suits.

12th. To colour Hats Green on the under side.

TAKE two pounds of fustick, chip it fine, put it into two gallons of soft water, boil it four hours in brass, keeping nearly the quantity of water: take out the chips, add two ounces of curkemy root, and one ounce of allum; boil this to three pints, brush this on the hats twice over, then add to one quart of this yellow liquor, three tea-spoonfuls of the indigo and vitriol, (as mentioned in a former receipt) this will make it green, brush this on the hat two or three times, leaving time between for the hat to be nearly dry.

13th. To Colour Feathers, &c. Black.

THIS is the most difficult colour to set. The feathers must lay in a preparatory liquor twelve hours; as follows— To each quart of water add one tea-spoonful of aquafortis, it must be kept hot the whole of the time: then, for three ounces of feathers, take two pounds of logwood chipped fine, and one pound of common sumac, put these into three gallons of water in an iron kettle, boil it four or five hours, take out the chips, and add two ounces of English nutgalls pounded fine; boil the three gallons to three quarts, then put in the feathers, let them be twelve hours; then take three ounces of copperas, and one ounce of verdigrease made fine, put them into half a pint of urine, and stir it on a moderate fire ten or twelve minutes; put this to the dye, it will set the colour; let them be in twelve hours more, then they must be washed or rinced perfectly clean. It is possible that hatters and others who deal in black, may find somethind in this to their advantage.

N. B. The preceding receipts for feathers, fur, &c. are intended for hatters as well as dyers.

14th. To Lacker Brass and Tin-Ware.

TAKE gum gamboge one ounce, make it fine, put it into four ounces spirits of wine, let it be kept warm four hours: the method of using it for small ware, such as buckles for harness, &c. put them on a piece of sheet iron, heat them hissing hot, then dip them in the lacker one at a time, as fast as you please. For large work, let the ware be heated, apply the lacker with a fine brush; it gives a most beautiful yellow.

15th. To soften Steel—for engraving, &c.

MAKE a very strong lie, of unslacked lime and white oak ashes, of each an equal quantity; put in the steel, let it lay fourteen days—it will be sosoft as easily to be cut with a knife.

16th. To make Oil-Cloth for Hats, Umbrellas, &c.

TAKE one pint of linseed oil, add one ounce spirits of wine, one ounce of litharge of gold, and one ounce of sugar of lead, simmer them together half an hour; take Persian or sarsnet, tack it within a frame, a common case knife is used in laying on the oil; twice going over is sufficient.

17th To make Oil-Cloth for Carpets.

To one gill of dissolved glue add one gill of honey, and one pint of water, simmer these together, stir in it five or six ounces of Spanish white, the cloth being tacked as above, rub this on till the pores are filled. If the paint be properly prepared, it will neither break nor peal off.

18th. The Chinese method for rendering Cloth water proof.

TAKE one ounce of white wax, (melted) add one quart of spirits of turpentine; when thoroughly mixed and cold, then dip the cloth into the liquid and hang it up to dry till it is thoroughly dry.

By the above cheap and easy method, muslin, as well as the strongest cloths, will be rendered quite impenetrable to the hardest rains; and that without the ingredients used either filling up the pores of the cloth or injuring, in the least, its texture, or damaging, at all, the most brilliant colours.

19th. To boil Oil for Painting.

To one gallon of oil, add one ounce of white vitriol, and an ounce of sugar of lead, a quarter at a time; boil one hour. Or this, put in four ounces of litharge of gold, one quarter of a pound of red lead, one quarter of a pound of sugar of lead and one ounce of rosin made fine; heat over a moderate fire (but not burn), stir it two hours; let stand and settle; turn it off with care, and leave the lees.

20th. To make Stone Colour.

TO fourteen pounds of white lead, add five pounds of yellow ochre, and one cunce of ivory black; you may vary the shades, by adding with the lead, stone yellow and vermillion, and mix it with oil to your liking.

21st. To make Pearl Colour.

To twelve pounds of white lead, add one pound of stone yellow, half an ounce of Prussian blue, and two ounces of white vitriol to dry the paint. Vitriol is used in all paints for drying.

22d. To make deep Blue.

TO three pounds of white lead, add one once of Prussian blue. You may make your colour light or dark, by varying your lead and blue.

23d. To make Sea green.

To two pounds of stone yellow, add one ounce of Prussian blue.

24th. Verdigrease Green.

TO one pound of verdigrease, add two ounces of white lead. Prime your work with white lead, and lamp black ground with oil in proper order.

25th. Orange Colour for Carpets.

TO four pounds of stone yellow, add two pounds of red lead.

26th. To Paint Flesh Colour or Peach Blow.

TAKE white and red lead, grind them together: you may make any shade you please by varying the red and white lead.

27th. To Paint a Red Brown.

TAKE two pounds of Spanish brown, and one pound of red lead, and grind them with oil.

28th. To Paint Black.

TAKE lamp-black, and a small quantity of Prussian blue, and grind them with oil.

29th. To Slack Verdigrease.

TAKE a kettle of hot wet sand, wrap four or five ounces of verdigrease in a cabbage leaf, put as many of those parcels in the sand as is convenient, leaving two or three inches between; let them be in four hours, keeping the sand hot. The verdigrease being thus slacked a man may grind three times the quantity in a day as of unslacked.

30th. To make Vermillion.

TAKE of quick-silver eighteen pounds, of flour of sulphur six pounds; melt the sulphur in an earthen pot, and pour in the quick-silver gradually, being also gently warmed, and stir them well together with the small end of a tobacco pipe. But if from the effervescence, on adding the latter quantity of quick-silver, they take fire, extinguish it by throwing a wet cloth (which should be had ready) over the vessel. When the mass is cold, powder it, so that the several parts may be well mixed together. But it is not necessary to reduce it, by nicer levigation, to an impalpable state. Having then prepared an oblong glass body, or sublimer, by coating it well with fire, lute over the whole surface of the glass, and working a proper rim of the same around it, by which it may be hung in a furnace, in such a manner that one half of it may be exposed to the fire, fix it in a proper furnace, and let the powdered mass be put into it, so as to nearly fill the part that is within the furnace, a piece of broken tile being laid over the mouth of the glass, Sublime, then, the contents, with as strong a heat as may be used without blowing the fumes of the vermilion out of the mouth of the sublimer. When the sublimation is over, which may be perceived by the abatement of the heat towards the top of the body, discontinue the fire; and after the body is cold, take it out of the furnace, and break it; then collect together all the parts of the sublimed cake, separating carefully from them any dross that may have been left at the bottom of the body, as also any lighter substance that may have been formed in the neck, and appears to be dissimilar to the rest. Levigate the more perfect part; and when reduced to a fine powder, it will be vermilion proper for use; but on the perfectness of the levigation depends, in a great degree, the brightness and goodness of the vermillion. In order, therefore, to perform this, it is necessary that two or three mills, of different closeness should be employed, and the last should be of steel, and set as finely as possible.

31st. Of Rose Lake, commonly called Rose Pink.

TAKE Brazil wood six pounds, or three pounds of Brazil and three pounds of peachy wood. Boil them an hour with three gallons of water, in which a quarter of a pound of allum is dissolved. Purify then the fluid by straining through flannel, and put back the wood into the boiler with the same quantity of allum, and proceed as before; repeating this a third time. Mix then the three quantities of tincture together, and evaporate them till only two quarts of fluid remain. Prepare in the mean time, eight pounds of chalk, by washing over; a pound of allum being put into the water used for that purpose, which, after the chalk is washed, must be poured off, and supplied by a fresh quantity, till the chalk be freed from the salt formed by the allum; after which, it must be dried to the consistence of stiff clay. The chalk and tincture, as above prepared, must be then well mixed together by grinding, and afterwards laid out to dry, where neither the sun nor cold air can reach it; though if it can be conveniently done, a gentle heat may be used.

The goodness of rose pink lies chiefly in the brightness of the colour and fineness of the substance; which last quality depends on the washing well the chalk. The more the hue of rose pink verges on the true crimson, that is to say, the less purple it is, the greater its value.

32d. For Prussian Blue.

TAKE of blood any quantity, and evaporate it to perfect dryness. Of this dry blood powdered take six pounds, of the best pearlash two pounds; mix them well together in a glass or stone mortar, and then put the mixed matter into large crucibles or earthen pots, and calcine it in a furnace, the top of the crucible or pot being covered with a tile, or other such convenient thing, but not luted. The calcination should be continued so long as any flame appears to issue from the matter, or rather till the flame becomes very slender and blue; for if the fire be very strong, a small flame would arise for a very long time, and a great part of the tinging matter would be dissipated and lost. When the matter has been sufficiently calcined, take the vessels which contain it out of the fire, and as quickly as possible throw it into two or three gallons of water; and as it soaks there, break it with a wooden spatula, that no lumps may remain; put them in a proper tin vessel, and boil it for the space of three quarters of an hour or more. Filter it while hot through paper, and pass some water through the filter when it is run dry, to wash out the remainder of the lixivium of the blood and pearlash: the earth remaining in the filter may be thrown away. In the mean time, dissolve of clean allum four pounds, and of green vitriol or copperas two pounds, in three gallons of water: add this solution gradually to the filtered lixivium, so long as any effervescence appears to arise on the mixture; but when no ebullition or ferment follows the admixture, cease to put in more. Let the mixture then stand at rest, and a green powder will be precipitated; from which, when it has thoroughly subsided, the clear part of the fluid must be poured off, and fresh water put in its place, and stirred well about with the green powder; and after a proper time of settling, this water must be poured off like the first. Take then of spirits of salt, double the weight of the green vitriol, which was contained in the quantity of solution of vitriol and allum added to the lixivium, which will soon turn the green matter to a blue colour; and after some time, add a proper quantity of water, and wash the colour in the same manner as has been directed for lake, &c. and when properly washed, proceed in the same manner to dry it in lumps of convenient size.

It is necessary, in all painting, that all paints when mixed together with the oil, to grind it till it is a perfect salve, so as when you rub it between your fingers you cannot feel any roughness with it, but feel perfectly smooth as oil; then it is ground fit for use—then add oil, and stir it together what is necessary, or according to your liking. Oil must be boiled in all painting.

33d. To lay Gold Leaf on Carved, or Moulding Work.

TAKE stone yellow, and white lead an equal quantity; grind it fine with old oil: brush this smooth over the work twice; let stand twenty-four hours, and then cut your leaf in proper form on a leather cushion with a sharp knife, take up your leaf on cotton-wool, and put it to your work;

a light brush over the work after the gold is on will add to its beauty.

34th. MEMOIR

On a method of Painting with Milk—by A. A. Cadet de Vaux. Member of the Academical Society of Sciences.—From the "Decade Philosophique."

I PUBLISHED in the "Feuille de Cultivateur," but at a time when the thoughts of every one were absorbed by the public misfortunes, a singular economical process for painting which the want of materials induced me to substitute instead of painting in distemper. Take skimmed milk, two quarts; fresh slacked lime, six ounces; oil of carraway, or linseed, or nut, four ounces; Spanish white, five ounces. Put the lime into a vessel of stone ware, and pour a sufficient quantity of milk to make a smooth mixture; then add the oil by degrees, stirring the mixture with a small wooden spatula; then add the remainder of the milk, and finally the Spanish white. Skimmed milk, in summer, is often curdled; but this is of no consequence to our purpose, as its fluidity is soon restored by its contact with lime. It is, however, absolutely necessary that it should not be sour; for in that case it would form with the lime a kind of calcareous acetite, susceptible of attracting moisture.

The lime is slacked by plunging it into water, drawing it out and leaving it to fall to pieces in the air. It is indifferent which of the three oils above-mentioned we use; however, for painting white, the oil of carraway is to be preferred, as it is colourless. For painting the ochres, the commonest lamp oil may be used. The oil, when mixed with the milk and lime, disappears; being entirely dissolved by the lime, with which it forms a calcareous soap. The Spanish white must be crumbled, and gently spread upon the surface of the liquid, which it gradually imbibes, and at last sinks; it must then be stirred with a stick. This paint is coloured like distemper, with charcoal levigated in water, yellow ochre, &c. It is used in the same manner as distemper. The quantity above mentioned is sufficient for painting the first layer of six toises, or fathoms.

One of the properties of my paint, which we may term milk distemper paint, is, that it will keep for whole months, and require neither lime nor fire, nor even manipulation; in ten minutes we may prepare enough of it to paint a whole house. One may sleep in a chamber the night after it has been painted. A single coating is sufficient for places that have already been painted. It is not necessary to lay on two, unless where grease spots repel the first coating; these should be removed by washing them with strong lime water or a lie of soap, or scraped off.

New wood requires two coatings. One coating is sufficient for a stair-case, passage, or ceiling. I have since given a far greater degree of solidity to this method of painting: for it has been my aim, not only to substitute it in the place of painting in distemper, but also of oil paint.

35th. Resinous Milk Paint.

FOR work out of doors I add to the proportions of the milk distemper painting, two ounces of slacked lime, two ounces of oil, and two ounces of white Burgundy pitch. The pitch is to be melted in oil by a gentle heat, and added to the smooth mixture of milk and oil. In cold weather the milk ought to be warmed to prevent its cooling the pitch too suddenly, and to facilitate its union with the milk of lime. This painting has some analogy with that known by the name of encaustic.

I have employed the resinous milk paint for outside window shutters, that had been previously painted with oil. The cheapness of the articles for this paint, makes it an important object for those people that have large wooden houses and fences.—An experiment has been made with this paint in this country, and it at present appears to answer perfectly the description of the inventor.

36th. An easy and cheap method to Stain Cherry a Mahogany Colour.

TAKE common whitewash of lime and water, white wash the wood, let it stand perhaps twenty-four hours, then rub it off, after polishing the wood apply linseed oil. By using a small piece of wood you may find when the colour suits.

37th. To make Cherrywood the Colour of Mahogany.

TAKE two ounces of Spanish brown, one of red lead, a quarter of an ounce of vermillion and half an ounce of spruce yellow, all ground fine and strained or sifted in clean water; mix it well and as thick as it will pour, then take a woollen cloth and dip it thereto, and rub your work, the more it is rubbed the better it will appear; wipe off the work, varnish and polish it.

38th. For a Dark Mahogany Colour.

TAKE two pounds of logwood chips, boil well till the strength is well out; take one pint of the liquor and put it in a bottle: then take two ounces of dragon's blood, make it fine and put it into a bottle, and add a pint of spirits of wine, which should be well steeped, when settled it is fit for use. First brush the wood with logwood liquor twice over, then with the dragon's blood and you will obtain the colour, then varnish or polish &c.

39th. To Stain White Wood the colour of Mahogony, or Black Walnut.

TAKE two pounds of logwood chips, boil three hours in water, have two quarts of liquor; then add to one gallon of water eight ounces of madder, let it stand twelve hours, keeping it warm, strain it off, then mix it with an equal quantity of the logwood liquor; it is applied as other stains;

when hot brushing it over, and letting it dry each time till it suits.

40th. To Stain any kind of White Wood a Dark Red, or Light Mahogany Colour.

TAKE two ounces of drugs called dragon's blood, make it fine ; put it into a pint of double-rectified spirits of wine ; let it stand six or seven days, shake it often, brush it on the wood till the shade suits.

41st. To make a Cherry Red, on White Wood of any kind.

TAKE of the brightest of logwood two pounds, boil out the strength, take out the chips, add a table spoonful of the rasping of gallant gill root, boil this one hour, strain the dye and boil it down to one quarter of the quantity ; brush it on the wood when hot, repeat it till the colour suits.

42d. The best Red Stain for Wood.

THIS is made by boiling two pounds of red-wood in two gallons of water, in the same manner as logwood, &c. is boiled; it is necessary to boil this in brass: when boiled down to a proper quantity, add one ounce of cochineal, and two ounces of cream of tartar made fine ; boil this half an hour, or till there is but one quart of the liquor ; apply it warm, and add a tea-spoonful of aquafortis.

43d. To make Green, on any kind of White Wood.

TAKE a yellow liquor as described in receipt 9th, add the vitriol and indigo, less or more, to make what shade is wanted. In all shades, it is necessary to repeat colouring three or four times, leaving time for the wood to dry betwixt each colouring ; the colour grows darker by standing.—The wood will not do to varnish short of six or seven days after staining.

44th. To Stain Green.

TAKE three ounces of verdigrease powdered ; put it in a glass bottle with a pint of good vinegar ; let stand two days with often shaking, and kept warm ; brush it on the wood till you obtain the colour required.

45th. To Stain a Light Orange Colour.

TAKE two ounces of carkeiny root pulverized and put in a glass bottle ; add to it a pint of spirits of wine, steep it twenty-four hours, shake it, and brush over till it pleases.

46th. To Stain Wood Black.

TAKE logwood liquor to give the ground work, then take two ounces of English nut galls made fine, put this in one quart of water, let it stand four days, shake it often, then brush it on, three or four times ; when almost dry, rub it over two or three times with strong copperas water ; like other stains it grows darker by standing

47th. Varnish for Wood either Stained or Painted.

THIS is made the same as in receipt 2d, except, instead of three ounces of gum shellack, take of it one ounce and a half, and one ounce and a half of gum sandrick ; it must be laid with a soft brush, and several times repeated ; after it has stood three or four days, take rotten stone made fine and sifted, mix it with water, then with a sponge or soft linen, rub it on till sufficiently polished.

N. B. If the varnish should be too thick, you may soften it with spirits of turpentine.

48th. Varnish.

AN excellent varnish has recently been discovered, made of one part of sandrac not pulverized, and two parts of spirits of wine, made cold and the solution promoted by frequent shaking.

AS the method of preparing Copal Varnish, is generally kept secret by those who are acquainted with it, and as a tradesman who is desirous of knowing it, is obliged to give some times an hundred dollars to another, to let him into the secret, and that upon condition of not imparting it to any body else—the following to some may not be unacceptable.

49th. To make Amber or Copal Varnish.

TAKE of white rosin four drachms, melt it over a fire in a glazed vessel, after which put in two ounces of the whitest amber you can get, finely powdered: this last is to be put in gradually, stirring it all the while with a small stick over a gentle fire, till it dissolves; pouring in now and then a little oil of turpentine, as you find it growing stiff, and continue this till your amber is melted. When the varnish has been thus made, pour it into a coarse linen bag, and press it between two hot boards of oak, or flat plates of iron. Great care must be taken in making the varnish, to not set the house on fire; for the vapour of the oil of turpentine will even take fire by heat.—If it should happen so to do, immediately cover the pot with a board or any thing that will suffocate it; by which means it will be put out.

50th. A Composition for giving a Beautiful Polish to Mahogany Furniture.

DISSOLVE bees-wax (equal parts) in oil of turpentine, until the mixture attain the consistency of paste.—After the wood intended to be polished is well cleansed, let it be thinly covered with the above composition, and well rubbed with a piece of oil carpet, until no dirt will adhere to its surface.

51st. To Prepare Glue for Use.

TAKE one ounce of isinglass, pounded fine, dissolve four ounces of good glue, in one quart of water and strain the isinglass with the glue into a small pot or vessel for that purpose, and put in half an ounce of allum, and boil them all together.

52d. To make an excellent Black Ink Powder, &c.

TAKE four ounces of nut galls powdered, two ounces of copperas calcined, half an ounce of allum, and half an ounce of gum arabic, all powdered, and kept close from the air.

To make Ink.—The above is sufficient to make three pints of ink. Take of rain or river water one quart, one pint of vinegar, or sour beer; put in the powder and shake well and kept warm, and frequently shook together.

53d. For Making Black Ink.

TAKE one quart of rain water, or water with ripe walnut shooks soaked in it, or the water soaked with oak saw dust; strain it off clean, then add one quarter of a pound of the best blue galls, two ounces of good copperas, and two ounces of gum arabic; put it in a bottle, stop tight, then shake it well every day till the ink is fit for use—but the older the better. The above articles must all be pulverized, before they are applied to the water.

To keep ink from freezing, apply a little spirits of any kind. To keep ink from moulding, apply a little salt therein.

54th. For Red Ink, &c.

TAKE three pints of sour beer (rather than vinegar) and four ounces of ground Brazil-wood; simmer them together for an hour; then strain off and bottle, well stopped, for use.

Or you may dissolve half an ounce of gum senegal, or arabic, in half a pint of water; then put in a penny worth of vermillion; put into a small earthen vessel and pour the gum water to it, and stir it well till it is well mixed together, and it willbe fit for use in twenty-four hours—but requires stirring before using. In the same manner and form, you may make any other coloured ink, as blue, green, yellow, purple, &c. For blue, use indigo or Prussian blue; for green, take verdigrease and vinegar; for yellow, use curkemy root and allum; for purple, use Brazil and logwoods, with allum and a little pearlash. It is necessary to steep these substitutes; strain and bottle off, add the gum and shake well together, and kept warm.

55th. Wonderful Cure of the Dropsy, by Dwarf Elder. From the Massachusetts Magazine.

SOME years ago, when the invalids from Chelsea were ordered to garrison at Portsmouth, there was among them a man greviously afflicted with the dropsy. He had already become so unwieldly as to be rendered incapable of doing any thing whatsoever, and was at last so corpulent that he could procure no clothes to fit him.

In this critical situation, an herb doctor chanced to come by, and seeing the man in that situation, said, 'Well, friend, what will you give me if I cure you?' The poor object, (who had already spent nearly the sum of forty pounds on the medical gentlemen, without relief) eyeing the doctor with a look of contempt, scarce vouchsafed to return him for answer, that his cure was impossible—and was preparing to leave him, when the doctor, stopping him, offered to cure him for a glass of rum. So extraordinary a proposal did not fail to awaken the attention of the man, who considered the extreme reasonableness of the demand, followed the doctor without speaking a word, into his laboratory, who taking out a bottle containing a black liquid presented it to his patient, telling him to drink it off that day, and when gone, to fetch his bottle for more.

Upon a curious examination of the contents of the bottle,

finding it not unpleasant to the taste, the dropsical man wisely concluded there could be no harm in it, if there was no good; and accordingly, taking the bottle, he at night (though despairing of success) ventured to drink before he went to bed about one half of the liquor, and immediately composed himself to rest. But he had scarcely been a quarter of an hour in bed, before the physic operated so strongly that he was obliged to get up and search for the necessary utensil. This was presently filled—upon which he groped about for the one belonging to his comrade, which, having found, he also filled— and (strange to tell) a tub which was in the next room, was nearly filled.—So strong an evacuation of urine produced, as we may well suppose, a very material alteration; for the next morning he was able to buckle his shoes, which he had not done for a long time.

He did not fail to call on the doctor for a fresh supply, which having obtained, he continued drinking at meals, &c. with such good effects, that he was completely cured in less than a week.

A matter of such importance could not fail to attract the attention of the whole regiment, among whom I chanced to be an eye witness of it; and asked him what the liquid was—he informed me that it was a decoction made of the leaves of dwarf elder. Yours, &c.

56th. Cure for the Dropsy.

TAKE a six quart jug of old hard cider, put therein a pint of mustard-seed, one double-handful of lignumvitæ shavings, one double-handful of horse reddish roots; let them simmer together, over a slow fire, forty-eight hours, when it will be fit for use. Take a tea-cup full of this liquid, three times a day; and it will work off the disorder by urine, without any trouble to the patient.

A most surprising instance of the efficacy of this simple medicine, has lately taken place in the case of Mr. WM. WRAY, of Lunenburg, who, from the worst state of the dropsy, has by it been restored to perfect health.

FROM A PHILADELPHIA PAPER.

The Editor having received from a friend the following Recipe for the Cure of a Cancer, is induced from the veracity of the writer, and the importance of such a remedy to many afflicted individuals, to lay it before the public.

57th. A Safe and Efficacious remedy for the Cancer.

TAKE the narrow leafed dock-root, and boil it in water till it be quite soft, then bathe the part affected in the decoction as hot as can be born three or four times a day; the root must then be marshed and applied as a poultice.

This root has proved an effectual cure in many instances. It was first introduced by an Indian woman, who came to the house of a person in the country who was much afflicted with a cancer in her mouth; the Indian perceiving someting was the matter, inquired what it was, and on being informed, said she would cure her. The woman consented to a trial, though with little hopes of success, having previously used many things without receiving any benefit. The Indian went out and soon returned with a root, which she boiled and applied as above, and in a short time a cure was effected. The Indian was very careful to conceal what these roots were, and refused giving any information respecting them; but happening one day to lay some of them down, and stepping out, the woman concealed one of the roots, which she planted, and soon discovered what it was. Not long after a person in that neighbourhood being afflicted with the same complaint in her face, she informed her of the remedy, and in two weeks she was cured. Some time after, a man was cured of a confirmed cancer upon the back of his hand; after suffering much, and being unable to get any rest, being told of this root, it was procured and prepared for him; he dipped his hand in the water as hot as he could bear it for some time; the root was then applied as a poultice, and that night he slept comfortably, and in two weeks his hand was entirely cured.

Daniel Brown's father, having had a cancer in his head, had it cut out, and apparently healed; but some of the roots remaining, it again broke out: his doctor then informed him that nothing more could be done, except burning it out with hot irons; this being too harsh a remedy to submit to, he was much discouraged. The dock root was soon after recommended, and it cured him in a short time.

In the beginning of the winter of 1798, a hard lump appeared in the middle of my under lip, and in a short time became sore: it continued in that situation till spring, when it increased and became painful: I then shewed it to a person of skill, and soon found he apprehended it to be cancerous; after two or three different applications, the complaint increased and spread rapidly. Lot Trip, having heard of my complaint, mentioned this root—I called on him to know the particulars of it; he gave me the necessary information: the root was procured, and used in the manner above-mentioned, taking a mouthful of water in which the roots were boiled, and let it drop over my lips as hot as I could bear it; this I did three or four times a day, and then kept the root to it a day and a night; and in two days the pain entirely left me and in two weeks it was cured.

58th. Remedy for Cancers.

BURN half a bushel or three pecks of green old field red oak bark to ashes; boil these ashes in three gallons of water until reduced to one; strain that one gallon off, and boil it away to a substance similar to butter-milk or cream; apply a small quantity on a piece of silk or lint to the cancer, but no larger than the place or part affected. I have known two plaisters to effect a cure, where the cancer lay in a proper position for the medicine immediately to penetrate to the roots of it; otherwise, it may take several plaisters, as the medicine must be repeated every two hours until the roots of the cancer are killed; then apply healing, salve, with a little mercurial ointment mixed thereon, and dress it twice a day until cured, which will certainly be the case in twenty or thirty days at farthest. I have known several persons entirely relieved by the above prescription: and one in particular, after two attempts by a skilful physician to remove the cancerous parts by exusion.

After being greatly alarmed myself from a cancer about three years ago, and having followed some time the directions of an experienced physician, I, contrary to his orders, and notwithstanding the fears of my family, happily applied two plaisters of the above medicine, and no symptoms of it have appeared since.

59th. Receipe for the Cure of the Hydrophobia, or the Bite of a Mad Dog.

[By a Physician of respectability in New-York.]

PLACE a blister on the wound immediately, the sooner the better; and even if this has been neglected till the wound has healed, it is necessary to apply it; also, apply blisters to the inside of the ancles, wrists, and between the shoulders of the patient, keeping two running at a time. Keep the patient in the free use of vinegar, either in food or drink; and if he has not got a tight room, make it so by hanging up blankets; then boil a quart or two of vinegar, place it in the room of the patient on a chafing-dish or kettle of coals, and let the patient continue in the room fifteen minutes at a time morning and evening, and often wet his ancles, feet and wrists with it.

Give him three or four doses of the following medicine in the course of three weeks, that is as often as one in five or six days:—Calomel eight grains, native cinnabar and salt of amber each four grains, to each dose, to be taken in the morning in molasses; also, give him a decotion of tea, made of sarsapharilla root and guiacum chips, (commonly called lignum vitæ dust). If the patient is actually labouring under the symptoms of the hydrophobia, give the several remedies more frequently; if soon after the bite as above. If the patient actually has the disorder, when first attended to, repeat the remedies until he recovers; if immediately after the bite, it will be necessary to attend him for three weeks, which generally, clears him from infection. His diet must be light and easy of digestion generally, though he may make a moderate use of animal food; but he must strictly avoid the use of spirituous liquors. The above is the general plan I follow. LOT TRIP.

60th. Cure for the Bite of a Mad Dog.

THE roots of elecampane, (the plant star-worth) pounded soft, boiled in new milk, and given plentifully to any thing that is bitten, during forty-eight hours, (keeping the subject from all other food havé been found an effectual remedy for this dreadful and frequently fatal malady.—*N. Y. Paper.*

61*st.* *Cure for the Bite of a Mad Dog.*

THE following remedy for the bite of a mad dog is recommended in the French papers :—A new laid egg is to be beaten up and put into a frying-pan, with oil of olives, cold drawn, and dressed, but not too dry. Into this is to be put a great quantity of powder of calcined oyster shells, which is to be sprinkled in such quantities as the mixture will absorb. This is to be given as a dose which is to be repeated for nine days fasting; and the wound is at the same time to be washed with salt water. The author of it professes to have tried it with repeated success, on man, dogs, and other animals.

FROM A CHARLESTON PAPER.

62*d.* *The Infallible Cure for the Dysentery.*

I HAVE been acquainted with it nearly forty years, and never knew it to fail. I have cured all that ever had it on my plantation, and myself several times. Not forty days past, I was afflicted with the dysentary, and cured myself with the receipt under written. About thirty years ago, I cured two persons in Charleston, who had been under the care of three physicians, and it had baffled their art and skill; yet this receipt cured them in a few days. The public may rely on the efficacy and infalibility of the receipt, viz —As soon as you find the flux is bad if possible before it comes to the dysentery, drink three or four tea-cupfuls of melted suet daily, say a cup full every three or four hours; let the food be the flour of well parched Indian corn made into a pap with new milk, and sweetened with loaf sugar; and let the drink be nothing else but a strong tea made with chiped logwood, or red oak bark, and sweetened with loaf sugar, though it will do without sweetening. When you find it is checked, make the tea weaker; should it stop too sudden, take a little salts. With the above simples, I can cure thousands without the loss of one. The cure will be effected in five, six or seven days.

63*d.* *Cure for the Dysentery.*

TAKE of the roots of the low-running blackberry vine, one large handful; make a strong tea of them in the same manner as you would make other tea, only let it stand on the coals a little longer.—Give two tea-cups full to an adult, and one to a child. After it has operated, give the patient a plenty of low balm tea, or cold water if preferred. Be careful when the appetite returns, to give them but a little to eat at a time, and that as often as the appetite calls, and no oftener. This blackberry root tea operates as a thorough but gentle purge in this complaint, and as soon as it operates;

it changes the nature of the stools; that is, instead of blood, &c. the stools will be of a greenish froth, and so will continue to be until they become natural.

64th. Cure for the Dysentery.

TAKE new churned butter without salt, and just skimming off the curdy part, when melted over a clear fire, give two spoonfuls of the clarified remainder, twice or thrice within a day, to the person so affected. This has never failed to make almost an instant cure.

65th. For the Dysentery & Colera, or Vomiting.

TAKE oil of pennyroyal, two drops to a table-spoonful of molasses, syrup or honey; after being well stirred up, let one tea-spoonful be administered every hour until it has the desired effect, which from experience, I can safely assure the public, will be found in every case of the above disorder, to be a speedy and certain cure. For a grown person, the dose may be doubled, and given in the same maner.

From an Old Lady.

66th. An Infallible Cure for the St. Anthony's Fire.

I AM neither physician, surgeon, apothecary nor nostrum-monger, (says a correspondent) but totally ignorant of the materia medica, except that I have swallowed large draughts of it, to cure me of painful returns of St. Anthony's Fire at spring and fall. In vain, alas! did I swallow; for the saint was constant in his visit at the accustomed time, notwithstanding the repeated prophecies of my doctor and apothecaries to the contrary. Fortunately for me, ten years since, I was favoured with a visit from a good lady, during the spring confinement, who told me, if I would at the time, take the elder tree blossoms and in the spring of the year, at each season, for a month, drink every morning fasting, half a pint of elder flower tea, and the same in the afternoon, that it would drown the saint. The next season of the elder tree blossoming, I followed her advice, as also the spring following, and have done so these nine years; since which time, the saint has not tormented me in the least. I have recommended this tea, from my experience of its efficacy, to ten of my fellow-sufferers since my own case, every one of whom has found it a specific remedy.

When the elder tree is in blossom, a sufficient quantity of the flowers should be gathered, in a dry day, and dried with great care for the spring use. The tea is made, by pouring a quart of boiling water on two handfuls of elder flowers, when green; a less quantity will do when dry. It may be drank hot or cold, as best suits the stomach. Each single blossom is not to be picked off, but the heads from the main stalks.

67th. For St. Anthony's Fire.

TAKE a purge; and anoint with the marrow of mutton.

68th. An admirable Recipe for a Consumption.

TAKE of Madeira, (or good generous mountain) wine,

two quarts; balsam of Gilead, two ounces; albanum in tears, (grossly powdered) two ounces, flowers of Benjamin half an ounce, let the mixture stand three or four days near the fire, frequently shaking; then add thereto, of Narbonne honey four ounces, extract of Canadian maiden hair eight ounces, shake the bottle well, and strain off the liquor The dose two tea-spoonfuls, to be taken once in four hours, in colt-foot tea or water, sweetened with capillaire.

N. B. The Canadian maiden-hair, which we now import from thence in great plenty is infinitely superior to that which grows in England. A strong infusion made of this herb, sweetened with honey or sugar candy, is the best ptisan which can possibly be drank by consumptive people, and will of itself cure any recent cough.

69th. Cure for the Heart Burn.

EAT two or three meats of peach-stones, of any kind of peach, and it will effect a cure immediately. Those which are dry are preferable.

FROM A VIRGINIA PAPER.

70th. Infallible and Effectual Cure for the Stone.

THROUGH the channel of your paper I request a publication of the following cure for the stone by dissolution. The gentleman by whose consent and desire, and upon whose authority the subsequent facts are offered to the public, is a Mr. Richard Major, of Loudon county, in this state, minister of the baptist society; a man of integrity, and much respected. Being in company with him a few days ago, I had the following relation from his own mouth:—

That having for a number of years been afflicted with that painful disease, he was at length informed that a certain physician, his name unknown, labouring under the same disease, being at Berkley spring, a negro man there proffered to cure him: This he at first disregarded, but expecting a speedy dissolution unless some aid could be obtained, afterwards sent for the negro, who agreed to cure him for three pounds. He accordingly undertook, and in a short time effectually eradicated the disorder. The physician then gave him his choice of freedom by purchase in lieu of the contract betwixt them, on condition he would disclose the means of the cure; to which the slave agreed. The receipt is the expressed juice of the horse-mint and red onions; one gill of each to be taken morning and evening till the complaint be removed. That he, Mr. Major, being urged to a trial of the above-mentioned remedy, submitted to it, though with some reluctance, as he conceived his term of life to be but short at most. Not having it in his power to procure green mint, so as to get the juice, he used instead thereof, a strong decoction of the dried herb: in other respects strictly adhereing to the prescription, which had the desired effect. He began the experiment in August, and within a week he had ocular demonstration of dissolution by the slightest touch of a particle that had passed from him, which continued so to do without pain or the least obstruction, until the stone was entirely dissolved, and the cure completely effected before the ensuing spring. That from the time the disorder began to yield as aforesaid, he daily recovered his health, strength and flesh, and was in as good plight as ever, age excepted, being at the time seventy two years of age, with an appearance corresponding with his own account; and as he farther said, without the slightest attack of the disorder from the time he began to use the above means of cure. This, at his request, is communicated to the public by

DANIEL ROBERDIEU.

71st. Indian Method of Curing Spitting of Blood.

[Communicated in a letter to the late Doctor Mead.]

THE following case is a very extraordinary one; but I know the gentleman to be a man of veracity, and had this account from his own mouth. He was of a thin, hectic constitution, and laboured under a troublesome pulmonary cough for some years; at last he was taken with an hæmoptœ, for which he had the best advice he could get in Maryland, but he grew rather worse under the care of two physicians who attended him for several months; and at last he was prevailed upon to put himself under the care of a negro fellow, who is the Ward of Maryland: for he has the reputation of performing some extraordinary cures, though nature has the chief claim to them: but indeed this was not the case here.—In short, he advised the gentleman to go into a warm bath twice a day, and sit up to his chin in it, for two or three minutes at a time, and as soon as he came out to dash cold water several times on his breast, and to wear flannel next his skin. This method soon relieved the gentleman; and when I left Maryland, which was about seven or eight years after the cure, he remained free from his hæmoptœ, eased very much of his cough, and went through a good deal of exercise.

72d. A Receipt for Bitters to prevent the Feve· and Ague, and all other Fall Fevers.

TAKE of common meadow calamus cut into small pieces, of rue, wormwood and camomile, or centaury, or hoar-hound, of each two ounces, add to them a quart of spring water, and take a wine glass full of it every morning fasting. This cheap and excellent infusion is far more effectual than raw spirits, in preventing fevers, and never subjects the person who uses it to an offensive breath, or to the danger of contracting a love for spirituous liquors.

73d. A certain Cure for Corns.

TAKE two ivy leaves and put them into vinegar for twenty-four hours; apply one of them to the corn, and when you find its virtue extracted, apply the other, and it will effectually and speedily remove the corn without the least pain.

74th. To make the most cheap and simple Electric Machine.

TAKE a piece of plank eighteen or twenty inches square, place two small posts at a distance that will take the length of a bottle that will hold perhaps a quart; the bottle must be round, and of flint glass, (they may be had at the apothecaries for 3s. or 3s. and 6d.) put in a hard wooden stopple, at the other end stick on a piece of hard wood with any glutinous matter, such as shoemaker's wax or the like; make a small hole in the center of this wood, and the stopple, to receive two points which come thro' the posts; thus the bottle being hung in a rolling position, let a band go round the neck, and be conveyed to a wheel, eight or nine inches over which turns with a crank. Then take an eight ounce vial, coat it inside and out with tin foil; this may be stuck on with stiff glue or candied oil; the vial must have a large nose, or it will be difficult to coat the inside; cork it tight, having a wire run through the middle of the cork with a common leaden bullet on the top; bind the wire so that the ball may come within half an inch of the cylinder or large bottle; place it in the center of the cylinder, then having a piece of deer-skin leather sewed up and stuffed in form of a pincusion, having amalgam rubbed on one side, hold it to the cylinder opposite to the ball; put the machine in motion, and the fire will collect and fill the small vial. To take a shock, hold the vial where it is coated with one hand, touch the ball with the other. If a number of persons wish to take a shock at once, the person at one end of the circle holds the vial, whilst that on the other touches the ball; the vial must not be coated within one inch of the top.

To make amalgam, take half an ounce of speltar, melt it, mix with it half an ounce of quick-silver; whilst warm, grind it to a powder. This machine is very useful where a stagnation of blood or any kind of numbness has taken place; for sudden pain, &c. The writer has reason to speak well of this machine, as it was one time the means of saving his life. It is sincerely wished that a physician or some other person would keep one in each town; the expence is no more than seven or eight shillings.

75th To Cure Children in the worst stage of Intoxication.

THE writer has twice known the instance of children, insensible of the effect of spirituous liquor, drinking to that degree that life was despaired of. On their being placed in a tub of warm water over their hips and a tea-kettle of cold water being poured on their head, they immediately recovered, and are now in perfect health. If this receipt may be the means of saving the life of but one child in the course of time, the writer will think himself richly paid for his trouble.

76th. Cure for the Ague.

DRINK the decoction, (that is the boiling of any herb) of camomile, and sweeten it with treacle; which drink when warm in bed- and sweat two hours. Or, to the wrists apply a mixture of rue, mustard, and chimney soot, by way of plaister.

77th. Cure for Almonds of the Ears fallen down.

TAKE a little bole armeniac in powder, and with it mix some Venice turpentine, and spread it on sheep's leather, as broad as a stay, and apply it under the throat from ear to ear.

78th. A Cure for Frost Bitten Feet.

TAKE the fat of a dung-hill fowl, and rub the place or places affected with it, morning and evening, over a warm fire; at the same time wrapping a piece of woolen cloth, well greased with the said fat, round the frost bitten parts. In two or three days they will feel no pain, and in five or six days will be quite cured.

Note.—If the inner bark of the elder, or the leaves of plantain, are first simmered in said, fat it will be the better.

79th. To Cure the Asthma, or shortness of Breath

TAKE a quart of aqua vitæ, one ounce of anniseed bruised, one ounce of liquorice sliced, and half a pound of stoned raisins; let them steep ten days in the above-mentioned, then pour it off into a bottle, with two spoonfuls of fine sugar, and stop it very close.

80th. To make Itch Ointment, a certain Cure for the Itch.

TAKE one ounce of gum arabic, dissolve in two gills of water; then take one pound of fresh butter, put it in with gum water, melt and try it together till the water is out; then let stand till no more than blood warm, then add two ounces of spirits of turpentine and two ounces of red precipitate, stir and mix them with the butter and gum, and box it up to keep it from the air, and fit for use. By carrying a box of it, it will be a preventitive against the disorder: it gives no disagreeable smell from the use of it. You may rub a little round your knees and elbows, and you may sleep with a person actually afflicted with the itch, without danger of catching the disorder: to cure the itch, take this ointment, rub off the pimples, warm the ointment if the weather is cold, and rub it over them, and continue it three times in a week, till the skin becomes smooth, which will be in a week or ten days; oint when going to bed: it is well to have clean linen, &c.

81*st. Cure for the Salt Rheum.*

TAKE one ounce of salts of tartar, dissolve it in twenty-six spoonfuls of fair water; then take one spoonful of pure lime juice and add a lump of loaf sugar as large as a walnut, let it dissolve; then add a spoonful of the tartar liquor dissolved as above, and give it the patient before eating, twice in twenty-four hours.

82*d. An effectual Cure for the Rheumatism.*

WHEN the patient is afflicted with this painful disease, take the tow of flax, and twist a large slack cord, and fasten it round the part affected and continue wearing it next the skin; it will effect a cure; have faith try it and see.

83*th. Good Cider as easily made as bad.*

TO make cider of early or late fruit, that will keep a length of time, without the trouble of frequent drawing off—Take the largest cask you have on your farm, from a barrel upwards; put a few sticks in the bottom, in the manner that house-wives set a lye cask, so as to raise a vacancy of two or three inches from the bottom of the cask; then lay over these sticks either a clean old blanket, or if that be not at hand, a quantity of swindling flax, so as to make a coat of about a quarter of an inch thick, then put in so much cleaned washed sand, from a beach or road, as will cover about six or eight inches in depth of your vessel; pass all your cider from the press through a table cloth, suspended by the corners, which will take out the pummice; and pour the liquor gently upon the sand, through which it must be suffered to filter gradually, and as it runs off by a tap inserted in your vessel, in the vacancy made by the sticks at the bottom, it will be found by this easy method, as clear cider can be expected by the most laborious process of refining; and all the mucilaginous matter, which causes the fermentation and souring of cider, will be separated so as to prevent that disagreeable consequence.

N. B. Other methods may be easily invented for passing the cider through the sand, which is the only essential part of the above process

84*th. Method of making Apple Brandy.*

The following receipt for making Apple Brandy, was communicated by Joseph Cooper, esq. of Gloucester county, New-Jersy, accompanied with a specimen of the liquor, made in the manner he represented The liquor is mild, mellow and pleasent; and greatly superior to apple spirits procured by the common process.

Put the cider, previous to distilling, into vessels free from must or smell, and keep it till in the state which is commonly called good, sound cider; but not till sour, as that lessens the quantity and injures the quality of the spirit. In the distillation, let it run perfectly cool from the worm, and in the first time of distilling, not longer than it will flash when cast on the still head and a lighted candle applied under it. In the second distillation, shift the vessel as soon as the spirit runs below proof, or has a disagreeable smell or taste, and put what runs after with the low wines. By this method, the spirit, if distilled from good cider, will take nearly or quite one third of its quantity to bring it to proof; for which purpose, take the last running from a cheese of good water cider, direct from the press, unfermented, and in forty-eight hours the spirit will be milder and better flavoured than in several years standing if manufactured in the common way. When the spirit is drawn off, which may be done in fi e or six days, there will be a jelly, at the bottom, which may be distilled again, or put into the best cider or used for making cider royal, it being better for the purpose that the clear spirit, as it will greatly facilitate in refining the liquor. JOSEPH COOPER.

85*th. A Receipt to make an excellent American Wine: communicated to the Burlington Society for promoting Agriculture and Domestic Manufactories; by Joseph Cooper, esq. of Gloucester county, New-Jersy.*

I PUT a quantity of the comb, from which the honey had been drained, into a tub to which I added a barrel of cider immediately from the press: This mixture was well stirred, and left to soak for one night. It was then strained, before a fermentation had taken place; and honey was added until the strength of the liquor was sufficient to bear an egg. It was then put into a barrel; and after the fermentation commenced, the cask was filled every day, for three or four days, that the filth might work out the bung hole. When the fermentation moderated, I put the bung in loosely, lest stopping it tight might cause the cask to burst. At the end of five or six weeks the liquor was drawn off into a tub, and the white of eight eggs, well beat up, with a pint of clean sand, were put into it—I then added a gallon of cider spirit; and after mixing the whole together, I returned it into the cask, which was well cleansed, bunged it tight and placed it in a proper situation for racking off when fine. In the month of April following, I drew it off into kegs, for use; and found it equal, in my opinion, to almost any foreign wine. In the opinion of many judges, it was superiour.

This success has induced me to repeat the experiment for three years; and I am persuaded, that by using the clean honey, instead of the comb, as above described, such an improvement might be made, as would enable the citizens of the U States to supply themselves with a truly federal and wholsome wine, which would not cost one quarter of a dollar per gallon, were all the ingredients procured at the market price; and would have this peculiar advantage over every other wine hitherto attempted in this country, that it contains no foreign mixture, but is made from ingredients produced on our own farms.

By order of the Society,

WM. COXE, jun. Secretary.

86*th. A Method of making Currant Wine, which had been practised by many and found to be genuine.*

[Extracted from the Transactions of the Philosophical Society of Philadelphia.]

GATHER your currants when full ripe; break them well in a tub or vat; press and measure your juice; add two thirds water, and to each gallon of mixture, (juice add water) put three pounds of muscovado sugar, the cleaner and drier the better; very coarse sugar, first clarified, will do equally as well: stir it well till the sugar is well dissolved, and then bung it up. Your juice should not stand over night if you can possibly help it, as it should not ferment before mixture. Observe that your cask be sweet and clean. Do not be prevailed on to add more than one third of juice, as above prescribed, for that would render it infallibly hard and unpleasent: nor yet a greater proportion of sugar, as it will certainly deprive it of its pure vinous taste.

OF MAKING SUNDRY SORTS OF BRITISH WINES.

87*th. Currant Wine.*

PICK the currants (when they are full ripe) clean from the stalks, then put them into an earthen vessel, and pour on them fair and clean hot water, that is, a quart of water to a gallon of currants; then bruise or marsh them together, and let them stand and ferment;

then cover them for twelve hours, strain them through fine linen into a large earthen crock, (as they say in Sussex) and then put the liquor into a cask, and thereto put a little ale-yeast; and when worked and settled, bottle it off. This is exceeding pleasant, and very wholesome for cooling the blood. In a weak's time it will be fit for bottling.

88*th. Artificial Claret*

TAKE six gallons of water, two gallons of the best cider, and thereto put eight pounds of the best Malaga raisins bruised; let them stand close covered in a warm place for two weeks, stirring them every two days well together; then press out the liquor into a vessel again, and add to it a quart of the juice of barberries, (which perhaps is best) to which put a pint of the juice of black cherries: work it up with mustard seed covered with bread past for three or four days, by the fire side; after which, let it stand a week; then bottle it off, and it will become near as good, if not so as to exceed, common claret.

89*th. Gooseberry Wine.*

The best way is to take for every three pounds of fruit, one pound of sugar, and a quart of fair water; boil the water very well, but you must put in the aforesaid quantity of sugar when it is boiled; bruise the fruit, and steep it twenty-four hours in the water; stir it some time, then strain it off, and put the sugar to it and let it stand in a runlet close stopped for a fortnight; then draw it off, and set it up in a cellar, and in two months, it will be fit to drink.

90*th. Raspberry Wine.*

TAKE the raspberries clear from the stalks; to a gallon of which put a bottle of white-wine, and let them infuse in an earthen vessel two or three days close covered; then bruise the berries in the wine, and strain them through fine linen gently; then let it simmer over a moderate fire; skim off the froth, and then strain it again, and, with a quarter of a pound of loaf sugar to a gallon, let it settle; then, in a half a pint of white wine boil an ounce of well scented cinnamon, and a little mace, and put the wine, strained from the spice, into it, and bottle it up.

91*st. Damson Wine.*

DRY the damsons in an oven after you have taken out your bread, then to every quart of damsons put three quarts of fair water, but first boil it very well; then put the water and damsons into a runlet with sugar; and having stood a time sufficient, bottle it of.

92*d. Wine of Grapes.*

WHEN they are full ripe, in a dry day, pick off those grapes that are ripest; and squeeze them in a vat or press made for that purpose, in which must be a fine canvass bag to contain the grapes, and when in the press do not squeeze them so hard as to break the seeds if you can help it; because the bruised seeds will give the wine a disagreeable taste: then strain it well, and let it settle on the lees in such a cask or vessel as you may draw it off without raising the bottom; then season a cask well with some scalding water, and dry it or sent it with a linen rag dipped in brimstone, by fixing it at the bogue, by the bung or cork; then put the wine into it, and stop it close for forty-eight hours; then give it vent at the bogue, with a hole made with a gimblet; in which put a peg or fawcet, that may be easily moved with the fingers; then, in about two days time, it will be fit for drinking, and prove almost as good as French wine.

93*d. Wine of Strawberries or Rasberries.*

MASH the berries, and put them into a linnen bag, as aforesaid for the grapes and squeeze them into a cask, and then let it work as in the aforesaid grape receipt, &c. In this manner may cherry wine be made; but then you must break the seeds, contrary to what was said before concerning the grapes.

94th. A short way for Cherry Wine.

SQUEEZE the juice of the cherries into a cask, and thereto put a small quantity of sugar, corresponding to the quantity of juice; and when stood a month, it will be a pleasant liquor.

95th. Black Cherry Wine.

IN the same manner, take one gallon or more of the juice of black cherries and keep it in a vessel close stopped till it works; and after it is fine, add an ounce of sugar to each quart, and a pint of white wine.

96th. Mead.

TAKE six gallons of water, and thereto put six quarts of honey, stirring it till the honey be thoroughly mixed; then set it over the fire, and when ready to boil, scum it very well: then put to it a quarter of an ounce of mace, and as much ginger, and half an ounce of nutmegs, some sweet marjoram thyme, and sweet briar, together a handful: then boil them in the liquid, then let it stand by till cold, and then barrel it up for use.

97th. To make Beer, without Malt.

TAKE thirteen gallons of water, boil and scum it, put two pounds of brown sugar and two pounds of treacle to it; boil them together half an hour, strain the liquor thro' a sieve, and put to it a penny worth or two of baum, when cold; work it a day and a night, then turn it: let it stand in the barrel a day and a night, then bottle it, and put into each bottle a tea-spoon full of brown sugar.

98th. To make good common Beer.

FOR a barrel of thirty two gallons take half a pound of hops, steep in four gallons of water two hours, strain off, then take one pound essence of spruce, and one gallon of molasses; mix them together, and put it in the barrel, and two cents worth of yeast, and fill with water: if it is summer it need not be warmed, but warm it in winter; when full shake it well, and stop it loosely and in four days it will be fit for bottling, and use.

99th For preserving Apples thro' the winter.

THE secret for preserving apples through the winter, in a sound state, is of no small importance. Some say that shutting them up in a tight cask is an effectual method, and it seems probable: for they soon rot in open air. But an easier method, and what has recommended itself to me by the experience of several years, is as follows:—

I gather them about noon, at the full of the moon, in the latter part of September or beginning of October. Then spread them in a chamber or garret, where they lie till about the last of November.—Then remove them into casks or boxes, in the cellar, out of the way of the frost; but I prefer a cool part of the cellar. With this management I find I can keep them till the last of May, so well that not one in fifty will rot.

100th. To pickle Cucumbers, green.

WASH them, and dry them in a cloth; then take water, vinegar, salt, fennel tops, some dill-tops, and a little mace; make it sharp enough for taste; then boil it awhile, then take it off and let it stand till cold; then put it in the cucumbers and stop the vessel close, and within a week they will be fit for use.

101st. To pickle French Beans.

TAKE them while young, and cut off the stalks, then take good vinegar and boil it with pepper and salt; season it to your palate, and let it stand till cold; then take the beans and put them into a stone jar, placing dill between the layers, and then put in the pickle, and cover them close for three weeks; then take the pickle and boil it again, and put it into the beans boiling hot; cover them close, and when cold they will be fit to eat.

Or, French beans may be pickled thus: Take your beans and string them, boil them tender, then take them off and let them stand till cold; then put them into pickle of vinegar, pepper, salt, cloves, mace, and a little ginger.

102*d. To pickle Walnuts, to eat like mangoes.*

TAKE green walnuts, before the shell has grown to any hardness in them; pick them from the stalk and put them in cold water, and set them on a gentle fire, till the outward skins begin to peel off; then, with a coarse cloth, wipe it off; then put them into a jar, and put water and salt therein, shifting it once a day for ten days, till the bitterness and discolouring of the water be gone; then take a good quantity of mustard seed, which beat up with vinegar, till it becomes coarse mustard; then take some clove of garlic, some ginger, and a little cloves and mace; make a hole in each nut, and put in a little of this; then take white-wine vinegar, and boil them together, which put to the nuts boiling hot, with some pepper, gi ger, cloves and mace, as also, some of the mustard seed and garlick, which keep close stopped for use.

103*d To pickle Mushrooms.*

FIRST blanch them over the crowns, and barb them beneath; then put them into a kettle of boiling water, then take them forth and let them drain; when they are cold, put them into your jar or glass, and put to them cloves, mace, ginger, nutmeg and whole-pepper; then take white-wine, a little vinegar, and salt; then pour the liquor into the mushrooms, and stop them close for use.

104*th. To Pickle Lemon and Orange Peel.*

BOIL them in vinegar and sugar, and put them into the same pickle: observe to cut them into small long thongs, the length of half the peel of your lemon; it ought to be boiled in water, before it is boiled in vinegar and sugar.

105*th. To Preserve Fruit green.*

TAKE pippins, apricots, pears, plumbs, or peaches, when they are green; scald them in hot water, and peel them; then put them into another water, not so hot as the first; then boil them very tender, and take the weight of them in sugar, and put to them as much water as will make a syrup to cover them; then boil the syrup till it be somewhat thick, and when cold put them together.

106*th. To Preserve Raspberries.*

TAKE good raspberries that are not too ripe, but very whole; take away the stalks, and put them into a flat bottomed earthen pan; boil sugar, and pour it over your raspberries, then let them stand to be cool; and when they are cold, pour them softly into your preserving kettle and let them boil till your syrup be boiled pretty thick; scum them very well in the boiling; this done, put them in pots, and when cold, cover them up close for use.

107*th. To Preserve Barberries.*

TAKE one pound of barberries picked from the stalks, put them in a pottle-pot, and set it in a brass pot full of hot water, and when they are stewed, strain them, and put to the barberries one and an half pounds of sugar, and to them put a pint of rosewater, and boil them a little; then take half a pound of the fairest clusters of barberries you can get, and dip them in the syrup whilst it is a boiling; then take the barberries out, and boil the syrup till it is thick, and when cold, put them in glasses with the syrup.

108*th. To Preserve Currants.*

LAY a layer of currants, and then a layer of sugar, and then boil them together as before prescribed for raspberries; scum them in boiling till the syrup is pretty thick; then take them off, and when they are pretty cold, put them in gallypots or glasses closely stopped.

109*th. To Preserve Walnuts green.*

BOIL the walnuts till the water tastes bitter, then take them off, and put them in cold water; peel off the bark, and weigh as much sugar as they weigh, and a little more water will then wet the sugar: set them on the fire, and when they boil up, take them off; let them stand two days, and then boil again.

110*th. To Preserve Cherries.*

FIRST take some of the worst cherries, and boil them in fair water, and when the liquor is well coloured, strain it; then take some of the best cherries, with their weight in beaten sugar; then lay one layer of sugar, and another of cherries, till all is laid in the preserving kettle; then pour a little liquor of the worst of cherries into it, and boil the cherries till they are well coloured: then take them up and boil the syrup till they will button on the side of a plate; and when they are cold, put them up in a glass close covered for use.

111*th. To Candy Cherries.*

TAKE cherries before they be full ripe, and take out the stones; then take clarified sugar boiled to a height, and pour it on them.

112*th. To Candy Pears, Plumbs, Apricots, &c.*

TAKE them, and give every one a cut half through; then cast sugar on them, and bake them in an oven, as hot as for manchet, close stopped; let them stand half an hour, then lay them one by one upon glass plates to dry, and they will appear very fine and clear: in this manner you may candy any other fruit.

113*th. Of Jellies.*

LET them be of apples, currents, raspberries, &c. Take out the clear liquor when squeezed, and boil it with sugar till it is as thick as a jelly. Then put into glasses, and cover it close.

114*th. A most excellent Method of making Butter, as now practised in England, which effectually prevents its changing and becoming rank.*

THE day before churning, scald the cream in a clean iron kettle, over a clear fire taking care that it does not boil over. As soon as it begins to boil, or is sufficiently scalded, strain it, when the particles of milk which tend to sour and change the butter are separated and left behind. Put the vessel in which it was strained in a tub of water, in a cellar, till next morning, when it will be ready for churning, and become butter in less than a quarter of the time required by the common method. It will also be hard, with a peculiar additional sweetness, and will not change. The labor in this way is less than the other, as the butter comes so much sooner, and saves so much labor in working out the buttermilk. By this method, good butter may be made in the hottest weather.

115*th. A method of Preserving Eggs.*

EGGS keep very well when you can exclude air; which is best done by placing a grate in any running water, and putting eggs, as the hens lay them, on the upper side of the grate, and there let them lie, covered with water, till you are going to use them, when you will find them as good as if they had been lain that day. This way answers much better than greasing; as sometimes one place is missed which spoils the whole egg : even those that are fresh never eat so well. In places where people are afraid their eggs will be stolen, they should make a chest with a number of slits in it, that the water may get in freely ; the top of which being above the water, may be locked down. Mill-dams are the most proper for these chests or grates.

N. B. The water must continually cover the eggs, or they will spoil.

116*th. To Cure Hams, as is practised in Virginia.*

TAKE six pounds of fine salt, three pounds of brown sugar, or three pints of molasses, and one pound of salt-petre powdered ; mix all these together, to serve for twenty-four hams : rub each ham well all over with this mixture, and pack them down in a cask or tub, and let them so remain for five or six days ; then turn them, and sprinkle some salt lightly over them, and so let them remain five or six days more, then add brine or pickle strong enough to bear an egg, and let them remain covered with it for a month, when they will be fit to smoak.

117*th. Another mode, equally as good and simple.*

TO four gallons of soft river water, add one pound of brown sugar, four ounces of salt petre, and eight pounds of coarse salt. Boil all these together, and carefully take off the scum as it rises ; when clear, let it remain till cold, then pour it over the meat till covered, and the quantity of pickle must be increased according to the quantity of meat ; the meat must not be pressed, but put lightly into a cask, and remain in for six or seven weeks, when it will be fit to smoke.

118*th. For a water to Destroy Bugs, Flies, Ants and other Insects, on tender plants.*

[Invented by C. Tatin, Seedsman and Florist at Paris.]

THE receipt for this valuable composition, and which obtained for the ingenious author a reward from the Bureau de Consultam, who desired it might be made as public as possible, is thus given in the celebrated Annales de Chimie —

Take of black soap four ounces, flour of sulphur four ounces, mushrooms of any kind four ounces, water wherein dung has been soaked, two gallons: and thus in proportion. Divide the water into equal parts; pour one part into a barrel, vat or any vessel of convenient size ; which should be used only for this purpose ; let the black soap be stirred in it till it is dissolved, and then add to it the mushrooms after they have been slightly bruised. Let the remaining half of the water be made to boil in a kettle : put the whole quantity of sulphur into a coarse linen cloth, tie it up with a thread in form of a parcel, and fasten it to a stone or other weight, to make it sink to the bottom. During twenty minutes, being the time that the boiling should continue, stir it well with a stick, and let the packet of sulphur be squeezed so as to make it yield to the water all its power and colour. The effect of the water is not rendered more powerful by increasing the quantity of ingredients. The water, when taken off the fire, is to be poured into the vessel, with the remaining water, where it is to be stirred a short time with a stick ; this stirring must be repeated every day, till the mixture becomes fœtid, (or putrid) in the highest degree. Experience shews, that the older and more fœtid the composition is, the more quick is its action. It is necessary to take care to stop the vessel well every time the mixture is stirred. When we wish to make use of this water, we need only sprinkle it on the plants, or plunge their branches into it : but the best manner of using it, is to eject it on them with a syringe, or squirt gun.

119th. *To Kill Lice on Cattle.*

TAKE a broad woollen list, as broad as your hand, that will go round about his neck; then wet the list well in train oil, and sew it about the beast's neck, and the lice will come to it, and it will kill them if there were ever so many; daub some about the breast in several places and they will come to it, and it will kill them. No flies in summer will come near any wound or sore, where this is applied, for it will kill them.

120th. *To Destroy Bugs, and rid Houses of them.*

TO remove these noisome and troublesome vermin, take oil of turpentine, wash over the walls and bedsteads with it, or particularly where there are any crevices, cracks or crannies, and they will die away, and the room, after some time using it, will no more be pestered with them.

The juice of wormwood and rue is very good to wash the bedsteads, crevices, or any place where you suppose they are, and if you would lie safe among thousands in a room, rince your sheets in water, wherein sassafras has been well steeped, and they will not enter upon them; or you may lay that wood in slices among your linen, and it will have the same effect. Keep your rooms airy and clean always.

AGRICULTURAL.

To Farmers.

121st *An easy method to preserve Wheat and Rye from the Weavil.*

AS you stack wheat, on every two or three layers of sheaves, spread some elder leaves and branches. This was communicated to me by a farmer, who tried the experiment with success last year. The same informant adds, that he has read in history, that the same remedy has been applied in Europe, when they have occasion to lay up a seven year's store, &c. As the remedy is easy, it is to be hoped that farmers will avail themselves of the advantage. Exporters of flour from the states have nothing so much to fear. Inspectors of flour ought to be guarded against this evil; no such flour ought to be suffered to leave the states. The credit of our flour abroad depends on the inspectors.

N. B. Lime, applied as above, will produce the same effect.

122d. *To preserve Indian Corn from Birds, &c.*

TO prevent your Indian corn when planted, from being taken up by birds or destroyed by worms or insects, take about one pint of tar to a bushel of seed corn, and in the like proportion for a greater quantity, and stir it well together till every grain receives a part of the tar. This will effectually answer the purpose required.

123d. *For Inoculating Fruit Trees.*

AUGUST and September are the proper months to inoculate or bud most kinds of fruit trees; an operation that every landholder should have some knowledge of. When a tree has finished its growth for the year, a bud is formed at the very tip or end of the twig; which denotes that it is in a proper state to bud or inoculate. Some trees are indeed an exception, as they continue growing almost the whole season, and may be budded through all July and August.

With a sharp knife, slit the bark of any twig not more than half an inch thick, and not less than a quarter of an inch. Carefully cut through the bark, but not to wound the wood under it. Let the slit be rather more than an inch long. In like manner cut half an inch long across this slit, at the bottom, so that the two cuts through the bark will resemble a ⊥ bottom upwards. Then take a bud of the fruit you wish to propagate, with its bark near an inch long, taking care to loosen it from the woody part of the stem, so as to put it off from your thumb and finger, separating the bark and the eye under the bud from the wood. If the eye is left on the wood, you must throw by the bud and take another. Then insert the bud under the ⊥, before described, and bind it down with woollen strings, or well soaked strips of bark of bass wood, leaving the eye of the bud to the air. In two or three weeks, the bud will unite with the stalks, when the string must be loosened. The stocks, may be cut away the next spring. This method is on many accounts better than grafting. It gives the farmer another chance, provided his grafts fail in the spring. Stone fruits succeed only or best with inoculation. Small twigs, too small for common grafting, answer well—and above all in this way, very little injury is done to the stock. In a fruit country, this method ought to be well understood. A correspondent says, that cow-dung, with the addition of a very little salt, is a good plaister for the wounds of fruit trees. When large limbs are cut of, the stumps should be covered to keep out the air. Too much salt will spoil the tree.

124*th. To take a Film off a Horse's Eye.*

BLACK Pepper, finely ground, and sifted thro' a piece of gauze; add thereto fine ground salt, of each as much as will lay on the point of a case knife, mixing them well together; then take as much dough as will thinly cover an ounce ball, make it flat, place the pepper and salt thereon, and roll them up, making the same about the size of an ounce ball; then put it as low down as possible in the off ear fastening the ear so as to prevent its falling out. The above takes off the worst of films, and no way injures the horse. This receipt has been used many years in this place with the greatest success.

125*th. A Cure for Sheep-Biting.*

AN intelligent farmer in New Jersey seized a dog which often worried and bit his sheep. He tied the leg of the dog by a tether to the leg of a strong active ram, and placed them on the top of a hill. The ram immediately began to kick and butt the dog, who after a little snapping, attempted to fly. The tether held him, so that the ram easily overtook, kicked and butted him. After a short time the ram, excited to exertion, raced down the hill, and forced the dog after him. When the dog was so punished as not to forget it, he was let loose, and would never touch a sheep afterwards.

126*th. An easy and sure Method to find due North and South.*

TAKE a smooth piece of board, draw on it four, five or six circles, fasten it on the top of a post, stick a pin in the centre which the circles are drawn on within each other; observe in the forenoon on which circle the shadow of the head of the pin strikes, and make a mark; then in the afternoon observe when it strikes on the other side of the same circle; then find the centre on the circle, then strike a line from one to the other, which cannot fail of being north and south.

FINIS.

Appendix A

Common Names of Chemicals Used in Dyeing**

Reprinted from Natural Dyes and Home Dyeing *by Rita J. Adrosko (Dover Publications, Inc., New York, 1971).*

Alum	Potassium aluminum sulfate	$KAl(SO_4)_2 \cdot 12\ H_2O$
Aqua ammonia	Ammonium hydroxide solution	NH_4OH
Aqua fortis	Nitric acid	HNO_3
Aqua regia	Mixture of HCl and HNO_3	$HCl + HNO_3$
Argol (Argal or Argil)	Crude potassium bitartrate, red or white, depending on whether it is deposited from red or white grapes	—
Bleaching powder	Calcium hypochlorite	$CaOCl_2$
Blue stone	Blue vitriol (below)	—
Blue vitriol	Hydrated copper sulfate	$CuSO_4 \cdot 5\ H_2O$
Borax	Hydrated sodium tetraborate	$Na_2B_4O_7 \cdot 10\ H_2O$
Brimstone	Sulfur	S
Caustic potash	Potassium hydroxide	KOH
Caustic soda	Sodium hydroxide	NaOH
Chalk	Calcium carbonate	$CaCO_3$
Chrome mordant	Potassium dichromate	$K_2Cr_2O_7$
Chrome yellow	Lead chromate	$PbCrO_4$
Cinnabar	Mercuric sulfide	HgS
Copperas	Hydrated ferrous sulfate	$FeSO_4 \cdot 7\ H_2O$
Cream of tartar	Potassium acid tartrate	$KHC_4H_4O_6$
Fuller's earth	Hydrated magnesium and aluminum silicates	—
Glycerine	Glycerol	$C_3H_5(OH)_3$
Green vitriol	Copperas (above)	—
Javelle water	Sodium hypochlorite solution	NaOCl
Lime water	Water solution of calcium hydroxide	$Ca(OH)_2 \cdot H_2O$
Lye	Caustic soda (above)	—
Marine acid	Muriatic acid (below)	—
Milk of lime	Calcium hydroxide suspended in water	$Ca(OH)_2$
Muriatic acid	Hydrochloric acid	HCl
Nitre	Potassium nitrate	KNO_3
Oil of vitriol	Concentrated sulfuric acid	H_2SO_4
Orpiment	Arsenic trisulfide	As_2S_3
Pearl ash	Purified potash (below)	K_2CO_3

**Chief reference: Francis M. Turner, ed., *The condensed chemical dictionary*, 2nd ed., rev. New York: The Chemical Catalog Co., Inc., 1930.

Peroxide	Hydrogen peroxide	H_2O_2
Potash	Potassium carbonate	K_2CO_3
Prussian blue	Ferric ferrocyanide	$Fe_4(Fe(CN)_6)_3$
Prussic acid	Hydrocyanic acid	HCN
Realgar	Arsenic monosulfide	AsS
Red orpiment	Arsenic bisulfide	As_2S_2
Sal ammoniac	Ammonium chloride	NH_4Cl
Sal soda	Hydrated sodium carbonate	$Na_2CO_3 \cdot 10\ H_2O$
Saleratus	Pearl ash overcharged with carbonic acid gas	—
Saltpetre	Nitre (above)	—
Sig	Urine, whose principal constituent is urea, a weakly basic nitrogenous compound	$CO(NH_2)_2$ (urea)
Slaked lime	Hydrated calcium hydroxide	$Ca(OH)_2$
Soda ash	Sodium carbonate	Na_2CO_3
Sour water	Dilute sulfuric acid	H_2SO_4
Spirit of salt	Muriatic acid (above)	—
Spirits of nitre	Dilute nitric acid	$HNO_3 \cdot H_2O$
Sugar of lead	Lead acetate	$Pb(C_2H_3O_2)_2 \cdot 3\ H_2O$
Tannic acid (tannin)	Gallotannic acid	$C_{14}H_{10}O_9$
Tartar	Argol (above)	—
Verdigris	Basic copper acetate	$CuO \cdot 2Cu(C_2H_3O_2)_2$
Vermillion	Cinnabar (above)	—
Vinegar	Dilute impure acetic acid	CH_3COOH
Vitriol	A sulfate, usually of iron or copper	—
Vitriolic acid	Oil of vitriol (above)	—
Washing soda	Sal soda (above)	—

Appendix B

Dyes Occasionally Mentioned in Dyers' Manuals Printed in America

Reprinted from Natural Dyes and Home Dyeing *by Rita J. Adrosko (Dover Publications, Inc., New York, 1971).*

Agaric		Black
Almond leaves		Yellow
Aloes		Purple
Artichokes		Green
Bear-berry	*Arctostaphylos uva-ursi*	Brown
Bindweed		Yellow-orange
Blackwood bark		Grey
Bloodroot	*Sanguinaria canadensis*	Red
Buckwheat		Blue
Chrysanthemum		Yellow
Convolvulus		Yellow-orange
Corn-marigold		Yellow
Dyers' savory	*Serratula tinctoria*	Yellow
Dyers' woodroof	*Asperula tinctoria*	Red
Ebony wood		Yellow-green
Fenugrec	*Trigonella foenum-graecum*	Yellow
Fenugreek	*Trigonella foenum-graecum*	Yellow
Hairy mistletoe		Yellow
Lady's bedstraw	*Galium tinctorium*	Red
Lombardy poplar	*Populus dilata* (= *P. nigra* var. *italica*)	Yellow
Magnolia	*Magnolia virginiana*	Yellow
Malacca bean	*Semecarpus anacardium*	Black
Mangrove bark	*Sweitenia mahogani*	Brown
Nephritic wood	Lignum peregrinum***	Yellow
Privet berries	*Ligustrum vulgare*	Green
Saffron	*Crocus sativus*	Yellow
Saw-wort	*Serratula tinctoria*	Yellow
Savory	*Serratula tinctoria*	Yellow
Sorrel	*Rumex acetosella*	Black
Sweet gale	*Myrica gale*	Yellow
Zant		Yellow
	Andromeda arborea (*A. Ferruginea* var. *arborescens*)	Black
	Cistus ledon	Yellow
	Coccus polonicus	Red
	*Mespilus canadensis*****	Red
	Virga aura canadensis***	Green

*** Not a botanical name.

****Contemporary botanical name, *Crataegus canadensis*.

A CATALOG OF SELECTED

DOVER BOOKS

IN ALL FIELDS OF INTEREST

A CATALOG OF SELECTED DOVER BOOKS IN ALL FIELDS OF INTEREST

DRAWINGS OF REMBRANDT, edited by Seymour Slive. Updated Lippmann, Hofstede de Groot edition, with definitive scholarly apparatus. All portraits, biblical sketches, landscapes, nudes. Oriental figures, classical studies, together with selection of work by followers. 550 illustrations. Total of 630pp. 9⅛ × 12¼.
21485-0, 21486-9 Pa., Two-vol. set $29.90

GHOST AND HORROR STORIES OF AMBROSE BIERCE, Ambrose Bierce. 24 tales vividly imagined, strangely prophetic, and decades ahead of their time in technical skill: "The Damned Thing," "An Inhabitant of Carcosa," "The Eyes of the Panther," "Moxon's Master," and 20 more. 199pp. 5⅜ × 8½. 20767-6 Pa. $3.95

ETHICAL WRITINGS OF MAIMONIDES, Maimonides. Most significant ethical works of great medieval sage, newly translated for utmost precision, readability. Laws Concerning Character Traits, Eight Chapters, more. 192pp. 5⅜ × 8½.
24522-5 Pa. $4.50

THE EXPLORATION OF THE COLORADO RIVER AND ITS CANYONS, J. W. Powell. Full text of Powell's 1,000-mile expedition down the fabled Colorado in 1869. Superb account of terrain, geology, vegetation, Indians, famine, mutiny, treacherous rapids, mighty canyons, during exploration of last unknown part of continental U.S. 400pp. 5⅜ × 8½. 20094-9 Pa. $7.95

HISTORY OF PHILOSOPHY, Julián Marías. Clearest one-volume history on the market. Every major philosopher and dozens of others, to Existentialism and later. 505pp. 5⅜ × 8½. 21739-6 Pa. $9.95

ALL ABOUT LIGHTNING, Martin A. Uman. Highly readable non-technical survey of nature and causes of lightning, thunderstorms, ball lightning, St. Elmo's Fire, much more. Illustrated. 192pp. 5⅜ × 8½. 25237-X Pa. $5.95

SAILING ALONE AROUND THE WORLD, Captain Joshua Slocum. First man to sail around the world, alone, in small boat. One of great feats of seamanship told in delightful manner. 67 illustrations. 294pp. 5⅜ × 8½. 20326-3 Pa. $4.95

LETTERS AND NOTES ON THE MANNERS, CUSTOMS AND CONDITIONS OF THE NORTH AMERICAN INDIANS, George Catlin. Classic account of life among Plains Indians: ceremonies, hunt, warfare, etc. 312 plates. 572pp. of text. 6⅛ × 9¼. 22118-0, 22119-9, Pa. Two-vol. set $17.90

ALASKA: The Harriman Expedition, 1899, John Burroughs, John Muir, et al. Informative, engrossing accounts of two-month, 9,000-mile expedition. Native peoples, wildlife, forests, geography, salmon industry, glaciers, more. Profusely illustrated. 240 black-and-white line drawings. 124 black-and-white photographs. 3 maps. Index. 576pp. 5⅜ × 8½. 25109-8 Pa. $11.95

THE BOOK OF BEASTS: Being a Translation from a Latin Bestiary of the Twelfth Century, T. H. White. Wonderful catalog real and fanciful beasts: manticore, griffin, phoenix, amphivius, jaculus, many more. White's witty erudite commentary on scientific, historical aspects. Fascinating glimpse of medieval mind. Illustrated. 296pp. 5⅜ × 8¼. (Available in U.S. only) 24609-4 Pa. $6.95

FRANK LLOYD WRIGHT: ARCHITECTURE AND NATURE With 160 Illustrations, Donald Hoffmann. Profusely illustrated study of influence of nature—especially prairie—on Wright's designs for Fallingwater, Robie House, Guggenheim Museum, other masterpieces. 96pp. 9¼ × 10¾. 25098-9 Pa. $7.95

FRANK LLOYD WRIGHT'S FALLINGWATER, Donald Hoffmann. Wright's famous waterfall house: planning and construction of organic idea. History of site, owners, Wright's personal involvement. Photographs of various stages of building. Preface by Edgar Kaufmann, Jr. 100 illustrations. 112pp. 9¼ × 10.
23671-4 Pa. $8.95

YEARS WITH FRANK LLOYD WRIGHT: Apprentice to Genius, Edgar Tafel. Insightful memoir by a former apprentice presents a revealing portrait of Wright the man, the inspired teacher, the greatest American architect. 372 black-and-white illustrations. Preface. Index. vi + 228pp. 8¼ × 11. 24801-1 Pa. $10.95

THE STORY OF KING ARTHUR AND HIS KNIGHTS, Howard Pyle. Enchanting version of King Arthur fable has delighted generations with imaginative narratives of exciting adventures and unforgettable illustrations by the author. 41 illustrations. xviii + 313pp. 6⅛ × 9¼. 21445-1 Pa. $6.95

THE GODS OF THE EGYPTIANS, E. A. Wallis Budge. Thorough coverage of numerous gods of ancient Egypt by foremost Egyptologist. Information on evolution of cults, rites and gods; the cult of Osiris; the Book of the Dead and its rites; the sacred animals and birds; Heaven and Hell; and more. 956pp. 6⅛ × 9¼.
22055-9, 22056-7 Pa., Two-vol. set $21.90

A THEOLOGICO-POLITICAL TREATISE, Benedict Spinoza. Also contains unfinished *Political Treatise*. Great classic on religious liberty, theory of government on common consent. R. Elwes translation. Total of 421pp. 5⅜ × 8½.
20249-6 Pa. $6.95

INCIDENTS OF TRAVEL IN CENTRAL AMERICA, CHIAPAS, AND YUCATAN, John L. Stephens. Almost single-handed discovery of Maya culture; exploration of ruined cities, monuments, temples; customs of Indians. 115 drawings. 892pp. 5⅜ × 8½. 22404-X, 22405-8 Pa., Two-vol. set $15.90

LOS CAPRICHOS, Francisco Goya. 80 plates of wild, grotesque monsters and caricatures. Prado manuscript included. 183pp. 6⅝ × 9⅝. 22384-1 Pa. $5.95

AUTOBIOGRAPHY: The Story of My Experiments with Truth, Mohandas K. Gandhi. Not hagiography, but Gandhi in his own words. Boyhood, legal studies, purification, the growth of the Satyagraha (nonviolent protest) movement. Critical, inspiring work of the man who freed India. 480pp. 5⅜ × 8½. (Available in U.S. only)
24593-4 Pa. $6.95

ILLUSTRATED DICTIONARY OF HISTORIC ARCHITECTURE, edited by Cyril M. Harris. Extraordinary compendium of clear, concise definitions for over 5,000 important architectural terms complemented by over 2,000 line drawings. Covers full spectrum of architecture from ancient ruins to 20th-century Modernism. Preface. 592pp. 7½ × 9⅝. 24444-X Pa. $15.95

THE NIGHT BEFORE CHRISTMAS, Clement Moore. Full text, and woodcuts from original 1848 book. Also critical, historical material. 19 illustrations. 40pp. 4⅝ × 6. 22797-9 Pa. $2.50

THE LESSON OF JAPANESE ARCHITECTURE: 165 Photographs, Jiro Harada. Memorable gallery of 165 photographs taken in the 1930's of exquisite Japanese homes of the well-to-do and historic buildings. 13 line diagrams. 192pp. 8⅜ × 11¼. 24778-3 Pa. $10.95

THE AUTOBIOGRAPHY OF CHARLES DARWIN AND SELECTED LETTERS, edited by Francis Darwin. The fascinating life of eccentric genius composed of an intimate memoir by Darwin (intended for his children); commentary by his son, Francis; hundreds of fragments from notebooks, journals, papers; and letters to and from Lyell, Hooker, Huxley, Wallace and Henslow. xi + 365pp. 5⅜ × 8. 20479-0 Pa. $6.95

WONDERS OF THE SKY: Observing Rainbows, Comets, Eclipses, the Stars and Other Phenomena, Fred Schaaf. Charming, easy-to-read poetic guide to all manner of celestial events visible to the naked eye. Mock suns, glories, Belt of Venus, more. Illustrated. 299pp. 5¼ × 8¼. 24402-4 Pa. $7.95

BURNHAM'S CELESTIAL HANDBOOK, Robert Burnham, Jr. Thorough guide to the stars beyond our solar system. Exhaustive treatment. Alphabetical by constellation: Andromeda to Cetus in Vol. 1; Chamaeleon to Orion in Vol. 2; and Pavo to Vulpecula in Vol. 3. Hundreds of illustrations. Index in Vol. 3. 2,000pp. 6⅛ × 9¼. 23567-X, 23568-8, 23673-0 Pa., Three-vol. set $38.85

STAR NAMES: Their Lore and Meaning, Richard Hinckley Allen. Fascinating history of names various cultures have given to constellations and literary and folkloristic uses that have been made of stars. Indexes to subjects. Arabic and Greek names. Biblical references. Bibliography. 563pp. 5⅜ × 8½. 21079-0 Pa. $8.95

THIRTY YEARS THAT SHOOK PHYSICS: The Story of Quantum Theory, George Gamow. Lucid, accessible introduction to influential theory of energy and matter. Careful explanations of Dirac's anti-particles, Bohr's model of the atom, much more. 12 plates. Numerous drawings. 240pp. 5⅜ × 8½. 24895-X Pa. $5.95

CHINESE DOMESTIC FURNITURE IN PHOTOGRAPHS AND MEASURED DRAWINGS, Gustav Ecke. A rare volume, now affordably priced for antique collectors, furniture buffs and art historians. Detailed review of styles ranging from early Shang to late Ming. Unabridged republication. 161 black-and-white drawings, photos. Total of 224pp. 8⅜ × 11¼. (Available in U.S. only) 25171-3 Pa. $13.95

VINCENT VAN GOGH: A Biography, Julius Meier-Graefe. Dynamic, penetrating study of artist's life, relationship with brother, Theo, painting techniques, travels, more. Readable, engrossing. 160pp. 5⅜ × 8½. (Available in U.S. only) 25253-1 Pa. $4.95

HOW TO WRITE, Gertrude Stein. Gertrude Stein claimed anyone could understand her unconventional writing—here are clues to help. Fascinating improvisations, language experiments, explanations illuminate Stein's craft and the art of writing. Total of 414pp. 4⅜ × 6⅜. 23144-5 Pa. $6.95

ADVENTURES AT SEA IN THE GREAT AGE OF SAIL: Five Firsthand Narratives, edited by Elliot Snow. Rare true accounts of exploration, whaling, shipwreck, fierce natives, trade, shipboard life, more. 33 illustrations. Introduction. 353pp. 5⅜ × 8½. 25177-2 Pa. $8.95

THE HERBAL OR GENERAL HISTORY OF PLANTS, John Gerard. Classic descriptions of about 2,850 plants—with over 2,700 illustrations—includes Latin and English names, physical descriptions, varieties, time and place of growth, more. 2,706 illustrations. xlv + 1,678pp. 8½ × 12¼. 23147-X Cloth. $75.00

DOROTHY AND THE WIZARD IN OZ, L. Frank Baum. Dorothy and the Wizard visit the center of the Earth, where people are vegetables, glass houses grow and Oz characters reappear. Classic sequel to *Wizard of Oz*. 256pp. 5⅜ × 8. 24714-7 Pa. $4.95

SONGS OF EXPERIENCE: Facsimile Reproduction with 26 Plates in Full Color, William Blake. This facsimile of Blake's original "Illuminated Book" reproduces 26 full-color plates from a rare 1826 edition. Includes "The Tyger," "London," "Holy Thursday," and other immortal poems. 26 color plates. Printed text of poems. 48pp. 5¼ × 7. 24636-1 Pa. $3.50

SONGS OF INNOCENCE, William Blake. The first and most popular of Blake's famous "Illuminated Books," in a facsimile edition reproducing all 31 brightly colored plates. Additional printed text of each poem. 64pp. 5¼ × 7. 22764-2 Pa. $3.50

PRECIOUS STONES, Max Bauer. Classic, thorough study of diamonds, rubies, emeralds, garnets, etc.: physical character, occurrence, properties, use, similar topics. 20 plates, 8 in color. 94 figures. 659pp. 6⅛ × 9¼. 21910-0, 21911-9 Pa., Two-vol. set $15.90

ENCYCLOPEDIA OF VICTORIAN NEEDLEWORK, S. F. A. Caulfeild and Blanche Saward. Full, precise descriptions of stitches, techniques for dozens of needlecrafts—most exhaustive reference of its kind. Over 800 figures. Total of 679pp. 8⅛ × 11. Two volumes. Vol. 1 22800-2 Pa. $11.95
Vol. 2 22801-0 Pa. $11.95

THE MARVELOUS LAND OF OZ, L. Frank Baum. Second Oz book, the Scarecrow and Tin Woodman are back with hero named Tip, Oz magic. 136 illustrations. 287pp. 5⅜ × 8½. 20692-0 Pa. $5.95

WILD FOWL DECOYS, Joel Barber. Basic book on the subject, by foremost authority and collector. Reveals history of decoy making and rigging, place in American culture, different kinds of decoys, how to make them, and how to use them. 140 plates. 156pp. 7⅞ × 10¾. 20011-6 Pa. $8.95

HISTORY OF LACE, Mrs. Bury Palliser. Definitive, profusely illustrated chronicle of lace from earliest times to late 19th century. Laces of Italy, Greece, England, France, Belgium, etc. Landmark of needlework scholarship. 266 illustrations. 672pp. 6⅛ × 9¼. 24742-2 Pa. $14.95

ILLUSTRATED GUIDE TO SHAKER FURNITURE, Robert Meader. All furniture and appurtenances, with much on unknown local styles. 235 photos. 146pp. 9 × 12. 22819-3 Pa. $8.95

WHALE SHIPS AND WHALING: A Pictorial Survey, George Francis Dow. Over 200 vintage engravings, drawings, photographs of barks, brigs, cutters, other vessels. Also harpoons, lances, whaling guns, many other artifacts. Comprehensive text by foremost authority. 207 black-and-white illustrations. 288pp. 6 × 9. 24808-9 Pa. $8.95

THE BERTRAMS, Anthony Trollope. Powerful portrayal of blind self-will and thwarted ambition includes one of Trollope's most heartrending love stories. 497pp. 5⅜ × 8½. 25119-5 Pa. $9.95

ADVENTURES WITH A HAND LENS, Richard Headstrom. Clearly written guide to observing and studying flowers and grasses, fish scales, moth and insect wings, egg cases, buds, feathers, seeds, leaf scars, moss, molds, ferns, common crystals, etc.—all with an ordinary, inexpensive magnifying glass. 209 exact line drawings aid in your discoveries. 220pp. 5⅜ × 8½. 23330-8 Pa. $4.95

RODIN ON ART AND ARTISTS, Auguste Rodin. Great sculptor's candid, wide-ranging comments on meaning of art; great artists; relation of sculpture to poetry, painting, music; philosophy of life, more. 76 superb black-and-white illustrations of Rodin's sculpture, drawings and prints. 119pp. 8⅜ × 11¼. 24487-3 Pa. $7.95

FIFTY CLASSIC FRENCH FILMS, 1912–1982: A Pictorial Record, Anthony Slide. Memorable stills from Grand Illusion, Beauty and the Beast, Hiroshima, Mon Amour, many more. Credits, plot synopses, reviews, etc. 160pp. 8¼ × 11. 25256-6 Pa. $11.95

THE PRINCIPLES OF PSYCHOLOGY, William James. Famous long course complete, unabridged. Stream of thought, time perception, memory, experimental methods; great work decades ahead of its time. 94 figures. 1,391pp. 5⅜ × 8½. 20381-6, 20382-4 Pa., Two-vol. set $23.90

BODIES IN A BOOKSHOP, R. T. Campbell. Challenging mystery of blackmail and murder with ingenious plot and superbly drawn characters. In the best tradition of British suspense fiction. 192pp. 5⅜ × 8½. 24720-1 Pa. $3.95

CALLAS: PORTRAIT OF A PRIMA DONNA, George Jellinek. Renowned commentator on the musical scene chronicles incredible career and life of the most controversial, fascinating, influential operatic personality of our time. 64 black-and-white photographs. 416pp. 5⅜ × 8¼. 25047-4 Pa. $8.95

GEOMETRY, RELATIVITY AND THE FOURTH DIMENSION, Rudolph Rucker. Exposition of fourth dimension, concepts of relativity as Flatland characters continue adventures. Popular, easily followed yet accurate, profound. 141 illustrations. 133pp. 5⅜ × 8½. 23400-2 Pa. $3.95

HOUSEHOLD STORIES BY THE BROTHERS GRIMM, with pictures by Walter Crane. 53 classic stories—Rumpelstiltskin, Rapunzel, Hansel and Gretel, the Fisherman and his Wife, Snow White, Tom Thumb, Sleeping Beauty, Cinderella, and so much more—lavishly illustrated with original 19th century drawings. 114 illustrations. x + 269pp. 5⅜ × 8½. 21080-4 Pa. $4.95

DEGAS: An Intimate Portrait, Ambroise Vollard. Charming, anecdotal memoir by famous art dealer of one of the greatest 19th-century French painters. 14 black-and-white illustrations. Introduction by Harold L. Van Doren. 96pp. 5⅜ × 8½.
25131-4 Pa. $4.95

PERSONAL NARRATIVE OF A PILGRIMAGE TO ALMANDINAH AND MECCAH, Richard Burton. Great travel classic by remarkably colorful personality. Burton, disguised as a Moroccan, visited sacred shrines of Islam, narrowly escaping death. 47 illustrations. 959pp. 5⅜ × 8½. 21217-3, 21218-1 Pa., Two-vol. set $19.90

PHRASE AND WORD ORIGINS, A. H. Holt. Entertaining, reliable, modern study of more than 1,200 colorful words, phrases, origins and histories. Much unexpected information. 254pp. 5⅜ × 8½. 20758-7 Pa. $5.95

THE RED THUMB MARK, R. Austin Freeman. In this first Dr. Thorndyke case, the great scientific detective draws fascinating conclusions from the nature of a single fingerprint. Exciting story, authentic science. 320pp. 5⅜ × 8½. (Available in U.S. only)
25210-8 Pa. $6.95

AN EGYPTIAN HIEROGLYPHIC DICTIONARY, E. A. Wallis Budge. Monumental work containing about 25,000 words or terms that occur in texts ranging from 3000 B.C. to 600 A.D. Each entry consists of a transliteration of the word, the word in hieroglyphs, and the meaning in English. 1,314pp. 6⅞ × 10.
23615-3, 23616-1 Pa., Two-vol. set $31.90

THE COMPLEAT STRATEGYST: Being a Primer on the Theory of Games of Strategy, J. D. Williams. Highly entertaining classic describes, with many illustrated examples, how to select best strategies in conflict situations. Prefaces. Appendices. xvi + 268pp. 5⅜ × 8½. 25101-2 Pa. $5.95

THE ROAD TO OZ, L. Frank Baum. Dorothy meets the Shaggy Man, little Button-Bright and the Rainbow's beautiful daughter in this delightful trip to the magical Land of Oz. 272pp. 5⅜ × 8. 25208-6 Pa. $5.95

POINT AND LINE TO PLANE, Wassily Kandinsky. Seminal exposition of role of point, line, other elements in non-objective painting. Essential to understanding 20th-century art. 127 illustrations. 192pp. 6½ × 9¼. 23808-3 Pa. $4.95

LADY ANNA, Anthony Trollope. Moving chronicle of Countess Lovel's bitter struggle to win for herself and daughter Anna their rightful rank and fortune—perhaps at cost of sanity itself. 384pp. 5⅜ × 8½. 24669-8 Pa. $8.95

EGYPTIAN MAGIC, E. A. Wallis Budge. Sums up all that is known about magic in Ancient Egypt: the role of magic in controlling the gods, powerful amulets that warded off evil spirits, scarabs of immortality, use of wax images, formulas and spells, the secret name, much more. 253pp. 5⅜ × 8½. 22681-6 Pa. $4.50

THE DANCE OF SIVA, Ananda Coomaraswamy. Preeminent authority unfolds the vast metaphysic of India: the revelation of her art, conception of the universe, social organization, etc. 27 reproductions of art masterpieces. 192pp. 5⅜ × 8½.
24817-8 Pa. $5.95

CHRISTMAS CUSTOMS AND TRADITIONS, Clement A. Miles. Origin, evolution, significance of religious, secular practices. Caroling, gifts, yule logs, much more. Full, scholarly yet fascinating; non-sectarian. 400pp. 5⅜ × 8½.
23354-5 Pa. $6.95

THE HUMAN FIGURE IN MOTION, Eadweard Muybridge. More than 4,500 stopped-action photos, in action series, showing undraped men, women, children jumping, lying down, throwing, sitting, wrestling, carrying, etc. 390pp. 7⅞ × 10⅝.
20204-6 Cloth. $21.95

THE MAN WHO WAS THURSDAY, Gilbert Keith Chesterton. Witty, fast-paced novel about a club of anarchists in turn-of-the-century London. Brilliant social, religious, philosophical speculations. 128pp. 5⅜ × 8½. 25121-7 Pa. $3.95

A CEZANNE SKETCHBOOK: Figures, Portraits, Landscapes and Still Lifes, Paul Cezanne. Great artist experiments with tonal effects, light, mass, other qualities in over 100 drawings. A revealing view of developing master painter, precursor of Cubism. 102 black-and-white illustrations. 144pp. 8¾ × 6⅜. 24790-2 Pa. $5.95

AN ENCYCLOPEDIA OF BATTLES: Accounts of Over 1,560 Battles from 1479 B.C. to the Present, David Eggenberger. Presents essential details of every major battle in recorded history, from the first battle of Megiddo in 1479 B.C. to Grenada in 1984. List of Battle Maps. New Appendix covering the years 1967–1984. Index. 99 illustrations. 544pp. 6½ × 9¼. 24913-1 Pa. $14.95

AN ETYMOLOGICAL DICTIONARY OF MODERN ENGLISH, Ernest Weekley. Richest, fullest work, by foremost British lexicographer. Detailed word histories. Inexhaustible. Total of 856pp. 6½ × 9¼.
21873-2, 21874-0 Pa., Two-vol. set $17.00

WEBSTER'S AMERICAN MILITARY BIOGRAPHIES, edited by Robert McHenry. Over 1,000 figures who shaped 3 centuries of American military history. Detailed biographies of Nathan Hale, Douglas MacArthur, Mary Hallaren, others. Chronologies of engagements, more. Introduction. Addenda. 1,033 entries in alphabetical order. xi + 548pp. 6½ × 9¼. (Available in U.S. only)
24758-9 Pa. $13.95

LIFE IN ANCIENT EGYPT, Adolf Erman. Detailed older account, with much not in more recent books: domestic life, religion, magic, medicine, commerce, and whatever else needed for complete picture. Many illustrations. 597pp. 5⅜ × 8½.
22632-8 Pa. $8.95

HISTORIC COSTUME IN PICTURES, Braun & Schneider. Over 1,450 costumed figures shown, covering a wide variety of peoples: kings, emperors, nobles, priests, servants, soldiers, scholars, townsfolk, peasants, merchants, courtiers, cavaliers, and more. 256pp. 8⅜ × 11¼. 23150-X Pa. $9.95

THE NOTEBOOKS OF LEONARDO DA VINCI, edited by J. P. Richter. Extracts from manuscripts reveal great genius; on painting, sculpture, anatomy, sciences, geography, etc. Both Italian and English. 186 ms. pages reproduced, plus 500 additional drawings, including studies for *Last Supper, Sforza* monument, etc. 860pp. 7⅞ × 10¾. (Available in U.S. only) 22572-0, 22573-9 Pa., Two-vol. set $31.90